DATE DUE			

THE CONSERVER SOCIETY

The text of this book is printed on 100% recycled paper.

The Conserver Society

*A Workable Alternative
for the Future*

Kimon Valaskakis

Peter S. Sindell

J. Graham Smith

Iris Fitzpatrick-Martin

With an Introduction
by Alexander King

HARPER & ROW, PUBLISHERS

New York, Hagerstown, San Francisco, London

FIRST EDITION

Designed by C. Linda Dingler

Library of Congress Cataloging in Publication Data

Main entry under title:

The Conserver society.
  Bibliography: p.
  Includes Index.
    1. Environmental policy—Addresses, essays, lectures.
I. Valaskakis, Kimon, 1941–
HC79.E5C6174   301.31   77-90868
ISBN 0-06-014489-0   79 80 81 82 83 10 9 8 7 6 5 4 3 2 1
ISBN 0-06-090671-5 pbk.   79 80 81 82 83 10 9 8 7 6 5 4 3 2 1

CONTENTS

PART III: CONSERVER SOCIETY TWO: THE AFFLUENT
STABLE STATE

PART IV: CONSERVER SOCIETY THREE: THE
BUDDHIST SCENARIO

PART V: THE SQUANDER SOCIETY OR CONSERVER
SOCIETY MINUS ONE

PART VI: ASSESSING THE OPTIONS

EPILOGUE

INTRODUCTION

Alexander King

Co-Founder, Club of Rome

Not very long ago, most people in most places looked forward
to the future essentially as a continuation of the present. Life
would be easier and more prosperous, thanks to the outpourings
of a benevolent technology, built on the findings of all-powerful
science, which would enable more and more sophisticated
goods to be produced more cheaply. Health prospects could be
expected to improve with advances in medical research; we
could expect to live longer, have more leisure and more means
to enjoy it. Society was seen as evolving in the direction of
greater equity, equality of education would eventually blur the
distinctions of class, and the grosser forms of poverty would
disappear as wealth trickled down.

Now we are not so sure. Violence, crime, and delinquency
are rife. Society creaks and often seems to be barely in control
as we lurch from one crisis to another—first, monetary upset,
then social, balance of payments problems, environmental dam-
age, ethnic and minority difficulties, unemployment, and the
rest, and then back to the next monetary flurry. The old economic tricks don't seem to work any more. Many people have
become alienated from society and, with a loss of faith in the
traditional religions and skepticism about the efficiency and
morality of the political process, there is a disastrous absence of
raison d'être. The future has suddenly become uncertain, and

technology with its unwanted side effects, although still the cornucopia, is seen to have many of the features of Pandora's Box.

Probing the future is very much à la mode. For a time 1984 was the symbolic year, then, as the end of the century began to loom on the horizon, the year 2000 became the landmark, although that too is now so close as to be of decreasing interest. The general malaise and uncertainty are felt to be no mere fin-de-siècle phenomenon; fin de millénium deserves a more sustained and detailed attention, and projections, trends, extrapolations, computerized analyses, and simple speculations have poured from academia and the think-tanks in wild contradiction one with the other, ranging from utopian optimism to apocalyptic doom, all to the confusion of the interested public.

In the aggregate, however, this rash of futurology has been extremely salutary; it has triggered (and its better offerings have given substance to) a debate which has spread the world over. To be successful, Cassandra always has to be willing to be proved wrong. If she is disbelieved and her warnings go unheeded, her prognostications may well come true, whereas if she is given credence, steps will be taken and policies changed so as to falsify them. This prophylactic objective is at the basis of today's more serious futures projections.

The grand debate centers around the question, how long can uncontrolled industrial growth continue? It was given a fresh impetus by the petroleum crisis, which taught us that oil is not forever, that energy is basic to all materials conversion and substitution, that it is subject to political as well as economic disruption, that the technological "fix" cannot be relied upon to give results sufficiently quickly, and that our complex social and technical structure is extremely fragile and vulnerable. The need for a basic change in the orientation of our societies, however, does not stem only from a realization that there are final material limits to growth. Many people, and especially the young, feel crushed by the material dominance in contemporary society, most strikingly within the urban environment. There must be an eventual saturation in the possession of goods and gadgets, which even the most valiant efforts of the persuasion industry and of planned obsolescence will not be able to overcome and which will, in the end, lead to increasing revulsion and rejection. Although nearly all governments still proclaim increased economic growth as a major objective and rely on it to get them out of the present economic and employment impasse, which they still, in their wishful thinking, insist is a temporary phenomenon, there is a widespread uneasiness that there are, indeed, limits to growth, social as well as material.

A further series of problems stems from the world population trends and from the disparities between the rich industrialized countries of the North and the poor countries of the Third World in the South. The demographic explosion, which involves doubling of the global numbers in a little over thirty years, is taking place mainly in the less-developed countries. Owing to recent population increases, the average age in these countries is already very low and is decreasing. During this period, the available work force in these countries will treble, whereas unemployment and underemployment are already rife. This phenomenon must inevitably have a major impact on the industrialized countries also. The huge numbers of new inhabitants of the world will make enormous demands on food, raw materials, and energy. Thus the lessened availability and resulting price increases will pose serious problems to such countries as Japan and those of Europe, which are resource poor, while for such nations as the United States, Canada, and Australia moral and material demands to raise food-production levels are unavoidable.

In the industrialized countries, population increase is slow and in some zero growth has already been achieved. This is beginning to raise serious problems. Industry in these countries will have to learn to cope with a constant and aging work force and with little expansion in the domestic markets. Educational systems must develop new policies to remain viable and innovative in the face of declining student enrollments and faculty numbers. National social-security schemes will have to seek greater efficiency and new methods if they are to avoid bankruptcy while financing the support of a growing proportion of retired citizens.

Thus, continued economic growth at levels to which we have become accustomed is likely to become difficult, if not impossible, despite the expanding markets in the Third World which increased populations there may provide. Yet there can be no question of establishing zero growth or "steady-state" economies the world over in the immediate future. The poverty and subsistence levels still existing in most of the world are such that a substantial increase in material standards, at least to provide the basic necessities of a decent life, will be demanded by the billions of the poor who are now aware, thanks to the media, of how the other half lives.

In the industrialized regions, decades of economic growth and material prosperity have generated demands for more and more goods and what we used to think of as luxuries have now become apparent necessities. Generalized expectations for an indefinite continuation of income and amenity increase will be difficult to meet, as will also demands for a more equitable distribution of wealth. Great problems of a social as well as of an economic nature are therefore already on the horizon, necessitating speedy consideration of alternative approaches to the evolution of society. For the industrialized societies the basic question now will be, how can we make much lower levels of material growth consistent with equity, personal liberty, cultural progress, and the satisfaction of physical and social needs? Even if levels of population and capital were to be held constant, there would still be freedom in deciding what the degree of social equity should be, the level of material well-being, the nature of innovation and research, the form of the political system, the extent and orientation of the social and service sectors, and the kind of employment, leisure, and cultural evolution.

The concept of the steady state is no novelty, but in a materialist society it is usually equated with stagnation and bleakness. In 1857, John Stuart Mill wrote, "It is scarcely necessary to remark that a stationary condition of capital and population implies no stationary state of human improvement. There would be as much scope as ever for all kinds of mental culture and moral and social progress; as much room for improving the art of living and more likelihood of its being improved." The problem thus resides in the value system of society and in the aspirations of its individuals. A high quality of life does not necessarily require material acquisition; it is essential that in working out the alternatives to material growth we keep in mind the positive advantages to happiness which each option might provide. Unless people everywhere appreciate the nature and extent of the dangers and difficulties which lie ahead and are also convinced that alternative and preferable lines of development exist, disillusion and social disruption are inevitable.

The tangled complexity of the world problématique cannot be tackled piecemeal. Changes in one component of society at a time seldom produce other than marginal amelioration and often even exacerbate the difficulties. For example, few of the problems are likely to find purely technical solutions, yet new technologies will have to be designed in keeping with changes in social values and goals, economic realities, laws, and institutional innovation. There will be no universal, easy, or quick fixes to the mass of problems which are rapidly making the social system unsustainable.

It is extremely difficult for governments, faced with immediate issues on the solution of which their survival depends, to take a firm grip on overall problems. Even where a few far-sighted political leaders realize the fundamental gravity of the situation, their people are insufficiently prepared to face the deep changes which are required, most of which will be unpopular, at least initially. Nor is it easy for a single country, however convinced it may be of the need to follow a new path of development, to act alone in a world which has become so interdependent. The machinery and procedures of government, created for earlier, simpler times, are incapable of responding quickly enough to cause fundamental change. The problems are mainly horizontal and interactive, whereas structures and policies are essentially constructed on a vertical, sector-by-sector basis. Likewise, and especially in the democracies with an electoral cycle of about four years, neither the administration in power nor the opposition is able to approach the voters on basic, long-term issues while immediate crisis interests dominate.

We therefore need an intensive period of debate and preparation, which cannot be too prolonged, oppressed as we are by the future, and inevitably much of this must be undertaken by unofficial groups, in universities and think-tanks, industrial groupings and trade unions, churches and schools. The task will be threefold: to deepen the understanding of the situation in the minds of the policy makers and the public at large; to explore in depth the nature and interconnection of the impending problems; and to present options and development possibilities. It was for such a purpose that the Club of Rome was created, and its value lies not so much in the facts, trend analyses, and forecasts which its reports have provided, as in the dialogue it has triggered and which has spread around the world. The present book is much to be welcomed for the same reason; it should become an important element in the great debate. To those of us who do not live in North America, it is interesting that the theme of the conserver, in contrast to the consumer, society should have been elaborated so articulately in this affluent continent. With their plentiful resources, their space, and their relatively small populations, the United States and Canada seem from outside to be exceptionally richly endowed. One realizes, of course, how empty is most of the space, particularly in Canada, that the minerals are exhaustible, that newspapers and advertising are eating up the forests, and that both countries are vulnerable to climatic change (as the winter of 1976–77 demonstrated!). Nevertheless, it is striking that the concept of the conserver society should be developed in such an environment.

The particular feature of the GAMMA approach is the examination of three separate scenarios of conservation—three different degrees of intensity and consequently of modification of or interference with existing life-styles. These range from one which the present population, conditioned by decades of materialistic dominance, should be able to accept without too much difficulty, to the Buddhist option, which would necessitate a complete change of values and living. In its recommendations the authors come down firmly in favor of the first, the only one which has any chance of acceptance at present, although only time will show whether it is in fact sufficient.

This scenario is based not only on increasing the efficiency of the production process to achieve a high degree of conservation of energy and materials, but also on an intelligent transformation of consumer habits.

It is interesting to note that the first of GAMMA's general recommendations to the Canadian Government is for the initiation of a nation-wide debate on the issues raised by the conserver society, and this must, as has already been stressed, be the first stage of any modification of the workings of society and any revision of national goals. It is greatly to be hoped that this recommendation will be quickly implemented.

The argumentation of the report is, of course, based on the realities of the American and Canadian scene, and its conclusions are intended essentially for the public of those countries.

The problems are, however, thoroughly general and apply to all countries. The report therefore should be useful in many other places, and the fact that the thinking originally came from Canada will give it credibility in a wide range of countries.

It is all too easy to label ideas such as those expressed in this book as naïve and utopian. To my mind they are much less naïve than the current belief that the existing approaches of consumerism and continuing high rates of physical growth can continue for a few more generations, if not indefinitely. Our present difficulties the world over are due in large measure to policies and practices of stimulating material growth without a commensurate effort to ensure that the existing quality of life can be maintained. An intensification of such policies to cure the present ills is a singularly naïve idea.

THE CONSERVER SOCIETY

1. THE BOOMERS, THE DOOMERS, AND THE CONSERVERS

Background to the GAMMA Report

From the first report to the Club of Rome, *The Limits to Growth,* we get a "zoom, doom, and gloom" view of the world. "Zoom" because all the key variables—population, garbage production, energy use, and pollution—zoom up in an alarmingly exponential way. "Doom" because the world system seems to be preset on an overshoot-and-collapse course. Everything will zoom until everything crashes at the same time, leaving humanity in one holy mess. Last, the message is "gloom" because some of the authors of that report and many of their disciples feel that nothing can be done to prevent catastrophe. Indeed, one of the earliest prophets of doom, the renowned demographer Paul Ehrlich, recently stated that "it is now too late to prevent a cataclysmic disaster."

In contrast, there are the "cornucopians," who believe in essence that if some is good, more is better, and most is best. Boasting many adherents, these "boomers" believe in continuing abundance—nay, in an infinity of resources available to us. This view is exemplified by the mainstream of the economics profession and finds strong support in Herman Kahn's 1976 book, *The Next Two Hundred Years*. Kahn shows the zoom curves almost miraculously bending and changing direction in the year 1976—coincidentally, of course, the year of the American Bicentennial. Indeed, the cover to the paperback edition graphically illustrates a two-stage human drama: 1776 to 1976 is a period of exponential curves, i.e., the zoom period; but around 1976 the curves subtly change and increase at a slower pace until 2176. (Mathematical minds will immediately recognize the S-curve.)

Presumably in the Kahn scenario the "good" zoom curves (increase in love, understanding, spiritual satisfaction, and so on) continue to accelerate. The boomers are inveterate optimists.

It is the peculiarity of our book that it is neither wholly optimistic nor wholly pessimistic. We agree (largely) with the "view of reality" of *Limits to Growth* and therefore recommend conservation, but we do not depict the conserver society as a "gloomy" prospect. Rather, we indicate that, as Cassius said, "the fault is not in our stars, But in ourselves"; we may decide our own future today.

This book describes the problématique of societal change in terms of five options: two are "cornucopian" and three relate to "conserver" societies. Problématique is not likely to be found in most English-language dictionaries, although it should be. In essence, a problématique, which originally was a French concept, is a structured hierarchy of questions and subquestions, not just a haphazard series. There is nothing as useless as a list. As one of our colleagues repeatedly notes, "Nature does not produce lists." Lists are confusing, boring, and often counterproductive. An unstructured shopping list, prepared stream-of-consciousness style, is the best way to spend an entire Saturday in a supermarket with minor results but major spending of time. The butter follows the apples which follow the mustard, the Ajax, and the chocolate chip cookies, and the hapless shopper turns around in circles trying to get the different items in haphazard order.

To avoid lists, a problématique structures a problem around

its central component parts and identifies all the subproblems emanating from them. It provides for the answering of the relevant questions in the right order. Without a problématique, more often than not much intellectual energy is devoted to answering either pseudo-problems or relevant problems in the wrong order.

The term "world-problématique" was made popular by the Club of Rome; yet the club's first report, *The Limits to Growth,* provided what in our view was not the best formulation of the problem. "Growth" in itself may be quite an innocuous term, but to speak of limits to growth may be to miss the point. "Throughput," on the other hand, is full of implications, some pleasant, some ominous. In essence "throughput" describes the process of transformation of "inputs" into "outputs," which corresponds to what we commonly call "production and consumption."

The fundamental problématique of this book focuses on the implications of "throughput." Different throughputs, however, lead to various growth or development options for society as a whole and for American and Canadian society in particular. In fact, we propose five throughput-centered growth options:

The status quo, characterized by the slogan "doing more with more," which we call CS_0 (Conserver Society Zero).

CS_1 (Conserver Society Model 1), which stresses growth with conservation or "doing more with less."

CS_2, the affluent or high-level stable state best characterized by the phrase "doing the same with less."

CS_3, the postindustrial conserver society, in which we would learn to "do less with less and do something else." CS_3 is the most radical of the conserver options and would require substantive value change.

CS_{-1}, the squander or anti-conserver society, whose creed is "do less with more."

The term "conserver society" was coined at the Science

Council of Canada in the early seventies. In its early years the phrase was used like Silly Putty or like Oscar Wilde's "sermon on the meaning of the manna in the wilderness." It could be brought out on any and all occasions, joyful or sad, festive or lugubrious, at weddings, anniversaries, christenings, or funerals. Like Cato's perennial exit line to the Roman Senate, "Carthage must be destroyed," it became fashionable in Canada to end many public speeches with the general admonition "We must move toward a conserver society." But beyond platitudes the concept remained vague.

In 1974, the GAMMA group, a think-tank at the University of Montreal and McGill University, was granted a contract by fourteen departments and agencies of the government of Canada to study the implications of a conserver society, that is, to make what was potentially a good idea into a policy option. The result was a four-volume study compiled from the technical papers written by fifteen experts each from a different discipline.* The results of the whole study were integrated in Volume One, and this book is directly derived from that volume of GAMMA's conserver society report.

The conserver society represents a much broader idea than conservation itself. Since Rachel Carson's *Silent Spring* there have been hundreds of good books on resource depletion, pollution abatement, energy conservation, and product recycling. There are now numerous manuals on solar energy, fuel economies, wind power, biomass agriculture, and we hope that there will be ever more such books. The conserver society, on the other hand, is a "package-deal" concept. It is a comprehensive (or near-comprehensive) set of scenarios reflecting the multidimensional nature of the human species.

All too often we tend to view problems in a partial way. Faced with an energy problem, we come out, as a society, with

---

* See "The Gamma Research Team" on page 278.

a few energy-conservative measures. Faced with inflation we set up anti-inflation policies. Faced with unemployment we dig out our anti-unemployment arsenal. We do all this without realizing that the solution to the energy problem may actually increase inflation, the anti-inflation program may worsen unemployment, and the anti-unemployment policies may result in increased waste in energy and materials.

The concept of the conserver society avoids such partial thinking by integrating as many aspects of the problématique as is humanly possible. The result is, we hope, alternative visions of the future, which in the final analysis reveal what the emerging science of futurology is really all about.

A final word of caution about data. A noted scientist is reported to have said: "If you ever want to impress an audience as being very profound, use big numbers. Start with the comment 'there are two billion stars in the galaxy.' You've got to be impressed. Nobody can count that high. Or say, 'there are a billion galaxies in the Universe.' If you are not impressed by that, there is something wrong with you."*

Some people's vision of the credibility or lack of credibility of an idea is inexorably linked with the presence or absence of big numbers in support of it. Yet numbers have a nasty way of being unreliable because there is always an "up-to-the-minute" new study with new numbers. The Meadows report based its credibility totally on the numbers game and exposed itself to scathing attacks. Instead of arguing for the basic existence of "limits to growth," the report gave a date for the Apocalypse—two hundred years from 1971. This was a bonanza for its detractors because such a prediction rested on highly questionable resource-reserve ratios, capital-output ratios, population projections, and the like. Viewed as a catalytic device to stir controversy, the Meadows report was immensely success-

*Herman Kahn, in "The New Class," an interview with Governor Jerry Brown of California in *CoEvolution Quarterly* (Spring 1977).

ful: it stimulated fertile scientific thinking. However, viewed as an argument to win a case it was badly flawed. There is a golden rule applicable to both human beings and computers: GIGO— garbage-in/garbage-out. A model is as good as its assumptions, and when these depend on questionable numbers the model itself becomes open to attack.

To avoid being one-upped by the "latest figures" (on "fossil-fuel reserves in the Alberta tar sands" or on "the number of milligrams of carbon dioxide in the air"), the arguments in this book are based on constants, not numerical variables. The numbers are presented as illustrative of a general direction of change. They are not presented as proofs that this book is holy writ, but neither can the "latest figures" claim any such eternal validity. The important thing is to separate the wheat from the chaff.

We try to avoid the tone of missionaries spreading a new creed. Although we end by advocating a conserver society as both a desirable and a feasible option for Canada and America, we strive to present the case in a reasoned manner, exposing the pros and cons of the status quo, the three conserver options, and the anti-conserver or squander scenario. The task of specialists is to present options, not to force a particular one down people's throats.

THE BIG ROCK CANDY MOUNTAIN

World View:
Cornucopia

Motto:
Do More with More

The Reform
of
Sammy Squander*

Being a fictional dramatization in nine parts of the plight, adventures, and tribulations of Sammy and his friends and the wisdom imparted to them by Madame Sosostris, famous clairvoyant. Any apparent resemblance to real persons is purely coincidental and most probably true.

*With the inspiration of T. S. Eliot, Jonathan Swift, William Shakespeare, Hesiod, Rachel Carson, and many others.

Sammy Squander Climbs the Big Rock Candy Mountain

"Relax! What with? How do you work the relaxo-vibrator on this couch?" These were the peevish questions Sammy Squander was directing at Dr. Fraudoong, who replied quietly and with his customary patience: "That *is* something we could use, Sammy, but right now we must get along as best we can. I've heard that some Indian swamis can relax even on a bed of nails. Just concentrate on letting go of each muscle, in turn, so that we can get at what's in your mind."

Sammy was reluctant to disclose that he had no idea where his muscles, if indeed he had any, were located. He was actually thinking that his Stratobird 2 + 2 was parked illegally outside the Sheraton, where he was lunching with friends in less than an hour. "A good-looking car seems to attract parking tickets like a Coke machine collects money," he thought. "Those Bolshie cops will be having a field day."

Sammy was getting into his usual frenetic state, constantly worried and on edge. He sometimes thought that life was just too much of a struggle—keeping up with the fashions in clothes, cars, bars, discothèques, skis . . . keeping appointments with the hairdresser, the masseur, the tailor, the tax accountant, the bank manager, the stockbroker, the consulting mechanic for his special-edition car . . . all that besides this twice-weekly visit to the shrink. Sammy had also been receiving the advice of a career consultant regularly over the five years since he had left college with a mediocre degree. He was sporadically and loosely involved in his father's "extremely successful" business, Squander Enterprises Unlimited, and often wondered how the employees found time to work.

Dr. Fraudoong, who seemed to want to earn his fee, was interrupting Sammy's routine reverie: "Just talk it out, Sammy. I have a new idea. We're always concentrating on your problems. Why don't we look at the positive side? I mean what you really get out of life, what you look forward to. Perhaps if you focus on some likable object in the room it will help you to concentrate on what you enjoy."

Sammy, obedient as ever, looked around the myriad of suggestive shapes decorating the consulting room and finally chose a red plastic triangle dangling from the stainless-steel light fitting. For the first time in these sessions, Sammy was inspired.

"That makes me think of a mountain," he said. "I'm on the mountain. I'm climbing up the slope, slowly, because I have all the wrong gear for this (I'm wearing my elevated shoes). But anyway, now I begin to see that it's not a normal mountain. It seems to be covered with all sorts of stuff... beautiful things. Wow... there's the new super electronic game set I'll collect from Ampower's this afternoon... and the new four-wall TV for a total experience that I'm thinking of getting. I think I can even see the new car I'll have next year—there it is, long, low, and beautiful. What a trip this is! There are other things, too. I can see a coffee fountain with whipped cream floating on the top... and different kinds of trees, shish-kebab trees, popsicle trees, money trees—there's even a machine with unlimited soft drinks in a hundred different flavors. Can you imagine that? It doesn't seem to be connected to anything but it works. It has mixers, too, for all the different drinks. I think I can see, a bit further up, one of the latest 18T Skimmers. You probably don't know what that is, but it's what I'll buy when I get a few thousand together—I really must have a cabin cruiser to join the Venturers Club."

"Keep going," urged Dr. Fraudoong. "What else do you see?"

"Well," Sammy sighed, "it's too much. There's everything... a tree with a few hundred light summer suits... another with just as many dress suits... a creases-away machine, shaving-cream dispensers, a scalp-massage set, electric dental floss... everything anybody could possibly want spread out all the way up the mountain, although I can't even see the top. There seems to be a lot of that old-fashioned rock candy I used to like as a kid—you know, the pink stuff with the white inside and something written right through? It says 'Souvenir from the Big Rock Candy Mountain.' That seems kind of familiar.... But do you know what the best thing is? There are dozens of Fionas [Fiona Fragment was Sammy's girl friend] all the same size, shape, same face but with different clothes, different hair styles, different makeup. There she is as a sporty-type redhead and over there she is as a groovy go-go girl dancing around on the mountain to the music. Oh, yeah, I didn't mention the music. It seems to be coming from the valley. I think maybe you'll know the song—it's some old thing my dad used to sing—it was a dream they had in the hard times. What was it called? The Depression.

> "O the buzzing of the bees in the cigarette trees
> The soda-water fountain
> Where the lemonade springs
> And the coconut sings
> On the Big Rock Candy Mountain!"

2. THE VALUES OF THE MASS-CONSUMPTION SOCIETY

All decisions, attitudes, and behavioral patterns which individuals or societies display depend on underlying systems of values. The process of valuation involves attaching a certain importance to X, a greater importance to Y, a lesser to Z, and so on and so forth. In some very fundamental sense our values are the ultimate criteria by which we judge our experiences, our perceptions, our behavior, and the behavior of others. To understand the modern mass-consumption society, it is therefore necessary to look into the intricate beliefs and values of the men and women who participate in it.

A value system can be said to be composed of at least three elements: beliefs, ideals, and preferences. A belief is a statement we hold to be true about the world (whether in fact it is or is not true is of course beside the point). An ideal is a judgment that a certain kind of behavior or a certain situation is "better" than another one. In fact, an ideal is a superlative and it describes not only the better but the best situations. Finally, a preference is an opinion stating that we like this and dislike that. A preference is distinct from an ideal. We might think it morally good to be chaste (an ideal) and yet have distinct preferences for nonchaste behavior. We might believe in the virtues of slimness (ideal) and proceed to gulp down a banana split de luxe.

The structure of beliefs, ideals, and preferences is, more often than not, implicit rather than explicit. We may know what our beliefs are at the conscious level but the structure of ideals and preferences is much more difficult to pinpoint, for the very simple reason that, insofar as the two may be in contradiction with each other, we prefer to hide them. Thus, it takes some prodding and some analysis—perhaps even some psychoanalysis—to get a person or a society to "confess" its values. Yet such a "confession" is necessary for an understanding of behavior.

At the root of the value system of the mass-consumption society are three important beliefs, which together may form what has been called a "paradigm" or world view.

First Belief: Happiness is Achieved Primarily Through the Accumulation of Things

In the depth of the Great Depression of the 1930s a popular song encapsulated the frustrations of an affluent society suddenly immobilized. "The Big Rock Candy Mountain" described a wonderland of plenty "where the roast duck springs and the coconut sings in the ice cold soda fountain."

The Big Rock Candy Mountain fantasy, legacy of the thirties, can be taken as an appropriate symbol of the current materialist paradigm: Happiness depends on the accumulation of material goods in ever-increasing numbers. The degree of our happiness is supposed to be exactly proportional to the number and monetary value of things we possess (i.e., the altitude reached on the Big Rock Candy Mountain).

The mountain of commodities even has its indicator: the gross national product. Growth of the GNP is interpreted as an increase in the standard of living (again the analogy of altitude) and therefore an increase in "happiness."

It is equally interesting to note that in micro-economic theory a person's well-being is measured by what are technically known as "indifference curves." These reflect levels of satisfaction and have been likened to "contour-maps on the mountain of happiness." Each contour line represents a certain altitude, and the steeper the climb the closer we can get to the elusive "bliss point"—in our terms the peak of the Big Rock Candy Mountain, where complete pleasure prevails and there is nothing left to do for an encore. It is the point of total saturation.

For good or ill, whether the idea is futile or fertile, it seems obvious that the current paradigm of the mass-consumption society is the Big Rock Candy Mountain. Lest you jump to the erroneous conclusion that the Big Rock Candy Mountain is the monopoly of the capitalist system, we should emphasize that Marxian orthodoxy is *dialectical materialism,* which totally espouses a materialist interpretation of history and the validity of the pursuit of a bigger and better material income. Where the Marxists differ from the capitalists is on the distribution of this material production. It is there that they introduce the notion of class struggle and make it an essential variable in social dynamics. But as far as the goals of Marxian society are concerned, they do not differ substantially from those of capitalist society. Both believe in the primacy and legitimacy of material accumulation as an end in itself.

Second Belief: Anthropocentrism

One of the legacies of the Judeo-Christian tradition is the belief in man's supremacy over nature. According to this doctrine, anthropocentrism: (a) Nature is created to satisfy our needs (i.e., nature is subservient to man). (b) Nature is endowed with unlimited resources. Therefore no threat of ultimate depletion of resources is to be taken seriously. (c) Nature is basically incompetent. It seems to do nothing right. Therefore, we must take it upon ourselves to transform, modify, and convert nature.

The idea that nature is created for man is well entrenched in the Judeo-Christian scriptures and surfaces in our language whenever we talk of "mastering the environment" or meeting the "challenge of nature." Unlike many Eastern religions, Western religions see man as separate from nature and not quite subject to its laws. This after all is the message of Genesis. The garden of Eden was there for Adam and Eve's good pleasure.

There is also an assumption that nature is evil and that we have to protect ourselves from its whims. Witness for instance the classic exchange between the two protagonists in Peter Weiss's play *Marat/Sade*.

MARAT: I read in your books, de Sade, in one of your immortal works, that the animating force of Nature is destruction and that our only instrument for measuring life is death.

SADE: Correct, Marat, but Man has given a false importance to death. Any animal, plant, or Man who dies adds to Nature's compost heap, becomes the manure without which nothing could grow, nothing could be created. Death is simply part of the process. Every death, even the cruellest death, drowns in the total indifference of Nature. Nature herself would watch unmoved if we destroyed the entire human race. I hate Nature, this passionless spectator, this unbreakable iceberg face that can bear everything, that goads us to greater and greater acts. But even though I hate this goddess I see the greatest acts in history have followed her laws. Nature tells Man to fight for his own happiness and if he must kill to gain it, why then the murder is natural. We must reproduce, we must destroy. The balance must be kept.*

In contrast is a North American Indian tribe's prayer to Nature:

O Grandmother Earth and Mother Earth we are of Earth and belong to You. O Mother Earth from whom we receive our food, You care for our growth as do our mothers. Every step that we take upon You should be done in a sacred manner. Each step should be a prayer.

DAKOTA PRAYER

*Peter Weiss, *The persecution and assassination of Jean Paul Marat as performed by the inmates of the asylum of Charenton under the direction of the Marquis de Sade*.

Not only is nature viewed as subservient and to some degree "evil" in the mass-consumption paradigm, but—strangely—also as bountiful and endowed with unlimited capacity to satisfy man's thirst. It is an immortal golden goose that cannot be destroyed. Because of this assumption, our treatment of the environment in industrial societies has been heavy-handed and parasitic—more so because of the belief that the host cannot be killed by the parasite—a proposition that is becoming less and less tenable today. The belief in the indestructibility of nature has allowed us to give ourselves license to abuse the environment wantonly and without regard to the consequences.

Third Belief: Incompetence of Nature

Finally, there is the curious assumption that nature is not only subservient and indestructible but also . . . incompetent. Industrial man sees himself as *homo faber,* man the doer, the transformer, the converter, forced to do, transform, and convert because nature makes such a mess of it. The dominant school of modern medicine, for instance, seems to be based on the assumption that surgery has to remove many of the body's organs when they malfunction and that an individual can survive properly only by supplementing his natural diet with dozens of pills and drugs. The scalpel-happy surgeons are quick to suggest operations to correct nature's incompetence, and pharmaceutical companies continuously promote their chemicals to counterbalance nature's excesses. Thus, we have "uppers" and "downers," stimulants, depressants, aphrodisiacs, appetite promoters, appetite quenchers . . .

Homo faber decides that a river is here to be dammed, a forest to be cut down, an animal to be hunted or domesticated, and a field to be landscaped. This belief has, of course, important implications because from it stems the idea of "transformation" or "throughput," which is at the core of an industrial system and which is discussed further on.

Among the ideals that characterize the mass-consumption society are two that have particular relevance in explaining our way of thinking: the work ethic and the growth ethic.

The work ethic is the legacy of early capitalism and the Protestant Reformation. In Max Weber's *The Protestant Ethic and the Spirit of Capitalism,* a relationship was shown between the capitalist spirit of accumulation and the Calvinist doctrine of salvation through hard work. The work ethic in the pure mass-consumption society advocated hard work leading to the building of a bigger and bigger mountain of commodities.

The work ethic that flourished in early American and pioneer days has now been somewhat deflected into a modern variant. The contemporary equivalent is the employment ethic, and there is a subtle distinction between the two. The work ethic of old enjoined the citizens to work hard and *productively.* The contemporary version enjoins society to provide a *job* for everyone, regardless of whether or not that particular job is in fact productive. As a result we tend to create jobs rather than get the job done, a misplacement of priorities that will be dealt with in detail in later chapters.

The employment ethic is also allied to the transformation bias of our mass-consumption economy. If nature is indeed incompetent and her work has to be constantly improved and modified by human hands, the source of economic value becomes labor-time. This idea is found in the classical works of Adam Smith, David Ricardo, and John Stuart Mill, but also in Marx, who, of course, makes labor the kingpin of the entire production process. Further, in modern economic accounting we speak of transformation as "value added," and one way of judging the economic performance of a nation is to measure the value added via transformation. The point is that all transformation is deemed useful, and the economic accounts are blind because they do not distinguish between productive and unproductive modifications.

The construction of an appliance which self-destructs after

one use leads to its replacement and to much "value added." The construction of a durable appliance, on the other hand, which survives repeated use represents less "value added." Curiously, our national accounts favor the former and have little good to say about the latter. This is all because our value system, as far as this aspect is concerned, is based on an activity model. The more activity the better. The more transformation the better, rather than production's being oriented toward fixed goals.

Not only does the mass-consumption society advocate employing everyone to busily transform our natural surroundings, whether for good or for ill, but the level of activity in doing so must be constantly increased. This is the growth ethic. The young culture of North America has always had a fetish about growth: industrial, muscular, personal. What is not growing is assumed to be dying. Therefore, the best way to guarantee vitality is by growth.

There is nothing necessarily wrong with the work ethic. As we will indicate in the following chapter, *some kinds* (but *not just any* kind) of growth may indeed be intimately related to the life cycle, be it of individuals, corporations, or societies.

One of the principal weaknesses of our mass-consumption society is its apparent inability to discriminate between types of growth. Instead, through a sort of social genetic code, cities, factories, products, and individuals reproduce themselves in ever-increasing numbers, creating dangerous exponential paths of expansion for certain sectors of society at the expense of others.

A spin-off of the growth- and work-ethic syndromes is the cult of newness. Unless something is "brand new" it is somewhat tainted, it has lost its virginity, it is to be avoided. It is well known that a car loses up to a third of its value as soon as the first owner inserts the ignition key. There is a stigma attached to a "used" car, a "used" appliance, or a used anything. It is a sign of poverty and therefore of failure. * Brand-newness means late-model cars, late-model appliances, late-model everything. That in its turn further incites to transformation and activity, which leads to producing and reproducing more late-model objects.

The other aspect of newness is private ownership. Unless an individual can have private possession of the things he or she wishes to consume, the benefits of newness disappear. What is the point of having the ''latest'' unless it confers upon the bearer some exclusivity? This leads right into the bias toward private property within the mass-consumption society, even if the thing owned is not really worth owning.

The temptation to buy things we rarely use just because we have been persuaded that we need them is yet a further element promoting mass production and therefore high levels of transformation.

William Leiss has shown in *The Limits to Satisfaction* that the mass-consumption society not only leads to the proliferation of needs but also to their fragmentation.† This is an important point. We have been taught to consider that we have very specific needs which can be satisfied only individually, leading to the fragmentation of needs, in turn promoting the fragmentation of commodities.

Witness, for instance, the typical modern man's overnight case. It contains hair spray, shaving cream, skin bracer, eye drops, toothpaste, mouthwash, dental floss, cologne . . . and we have, at this point, reached only the owner's neck.

Throughout history, society has attempted to satisfy the basic human needs. But how did we get into our present environmental crisis? Turning to its historical and cultural roots, we reach the heart of the issue: man's relationship to nature.

Lynn White and Raymond Dasmann have argued that particu-

* Antiques are an exception to this rule, of course.
†*The Limits to Satisfaction* (Toronto: University of Toronto Press, 1976).

lar, historically grounded attitudes toward nature have under-girded, conditioned, and reinforced the development of Western Europe and North America into urban, industrial societies with concomitant environmental destruction and degradation. White effectively chronicles the rise of an anthropocentric view of nature within the Judeo-Christian religious tradition, wherein man comes to share God's transcendence over nature. Chris-tianity shares with Judaism a creation myth in which man asserts dominance over the animal kingdom and in which "God planned all of the creation explicitly for man's benefit and rule; no item in the physical creation had any purpose save to serve man's purposes. By destroying the animist view of nature it became possible to exploit nature in a mood of indifference to the feelings of natural objects."

White claims that Judaism and Christianity also share a con-cept of time "as nonrepetitive and linear," which in our view clearly prepares the way for the nineteenth-century notion of progress (vital to the ideological underpinnings of our growth philosophy) and the paramount value accorded to technology and science. Reading White, one also sees an exploitative atti-tude toward nature rising during the medieval period, but even more striking perhaps is the emergence of an agricultural tech-nique, cross-ploughing, which cuts much more deeply into the soil. This forced the peasants to go beyond the familial level in pooling their oxen and labour. Despite such innovations, how-ever, peasants in Europe still were subject to the vagaries of weather, had a fairly dispersed settlement pattern, a deep sense of place, a close ongoing relationship to their natural surround-ings, and access only to relatively small-scale, low-impact tech-nology.

One can contrast the above with "the attitude of the transient exploiter" which charaterized the development of the western United States, according to Dasmann. He discusses "the bulldozer mentality" and states that "Americans are impatient with the slow processes of nature, with the normal events of biotic succession and change,'' preferring ''the simplicity of a machine to the intricacies of a biota.'' Canadians share this attitude. Americans and Canadians build towns, roads, and railroads on flood plains and in river canyons in defiance of common sense and natural law, while we let our Great Lakes die from pollution by mercury and PCBs. Recognizing this anthropocentric attitude is indispensable to an understanding of the mass-consumption society.

Economic kinds of values are dominant in the mass-consumption society. This can be seen most dramatically when we consider the human obsolescence so prevalent in an industrial society which fosters technological change and values it so highly. Jobs are lost as industries go into eclipse, factories go bankrupt, small stores are demolished in urban renewal projects, and so on.

Let us now examine the cultural dynamics of material consumption. The American anthropologist Jules Henry has developed three concepts which are vital to an understanding of the cultural dynamics of consumption: ''production-needs complementarity,'' ''coincidence,'' and ''the property ceiling.'' In small-scale traditional hunting and gathering and agricultural societies ''one does not produce what is not needed; and objects are made in the quantity and at the time required. Thus there is a congruence or complementarity between what is produced and what is desired and there is a coincidence with respect to timing.''

Underlying the economic activities in these societies, in contrast to our own, is the notion that there is a fixed, relatively unchanging configuration of wants and, therefore, level of production. The opposite cultural assumption about the nature of human nature pervades our society: human wants can be multiplied infinitely and unceasingly. Henry also notes that in almost all small-scale societies there are mechanisms for the redistribution of surpluses so that even when resources are abundant they reach large numbers of people and thus serve the public good rather than private profit.

But a production-needs complementarity is dangerous to a society driven by economic growth because, according to Henry, "it is only the deliberate creation of needs that permits the culture to continue." Many would argue that the economic system in our society exists merely to serve people's material needs, and that this in fact gives us a high level of welfare or happiness. But Ezra J. Mishan has demonstrated effectively how the unlimited marketing of goods leads to a cumulative reduction in the pleasure of people because of the conspicuous external diseconomies produced: noise, pollution, urban congestion. Many would argue that the market, rather than being a "want-satisfying mechanism," has become a "want-creating mechanism," principally through marketing and advertising.

Henry argues that the first two commandments of the mass-consumption society are: "1. Create more desire and 2. Thou shalt consume." Advertising pervades our life ubiquitously— on the bus, in the subway, in the car, at home, on the radio, on television, in the newspapers, in magazines, and sometimes even written in the sky. We are told to "buy now, pay later"; "fly now, pay later." Fifty cents of every consumer dollar is received by marketing intermediaries or used to pay manufacturers' marketing-related expenses.

3. THE MECHANICS OF MASS CONSUMPTION

The Notion of Throughput

The value system of the mass-consumption society is supported by an elaborate economic system, the principal elements of which are dominated by the idea of "throughput." All life processes share a basic unifying theme, the principle of transformation. To understand society as a living organism it is necessary to use biological metaphors.

There is a striking similarity between *metabolism* and the production-consumption continuum *throughput,* i.e., inputs becoming outputs (which may again become inputs) by a process of transformation. In biology, metabolism is the conversion of food by a living entity to the chemicals necessary for life. Metabolism involves two processes, catabolism and anabolism. In the catabolic phase, proteins and other nutrients from foreign bodies are broken down into their basic amino acids. In the anabolic phase they are reassembled into new structures to become part of the metabolizing entity's cells. The same is essentially true of any other throughput process. Inputs are broken down and reconstructed, which is exactly what happens in economic production and consumption. (See Figure 1.)

To attempt to distinguish production from consumption implies a heavily anthropocentric value judgment. Production is the creation of commodities, a commodity being an entity which can produce satisfaction in at least one person. Similarly con-

Figure 1
THE SIMILARITY BETWEEN
BIOLOGICAL METABOLISM AND THROUGHPUT

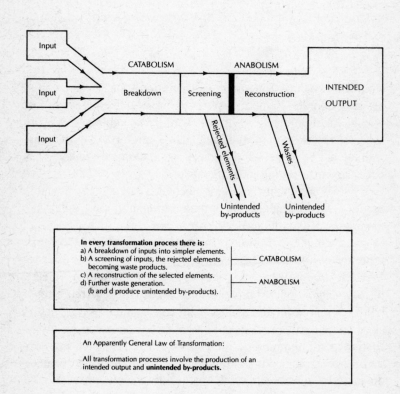

sumption is the destruction of a commodity, by removing from this entity (in its present form at least) the ability to produce satisfaction.

Obviously "production," "consumption," and "commodity" are defined above in solely human terms. When the hides of nine cows are used to produce the upholstery for the latest Rolls-Royce, the transformation of cow-into-leather is a productive one only from the human perspective; from the cows' point of view the transformation is destructive. This frivolous example serves to widen the rather narrow "world view" with which traditional economics has taken account of humanity's interaction with the environment. Particularly in the West since the inception of the Industrial Revolution, that interaction has consisted of radical structural change—clearing forests to be replaced by superhighways, building dams, landscaping, strip mining, urbanization. As long as such processes produced "commodities," their destructive side was ignored. Generally speaking, seeing them as purely productive was the result of taking a short-term perspective. To go back to the Rolls-Royce example, if a shortage of milk was foreseen, the cow-into-leather transformation would be destructive not only from the cows' point of view but also from a longer-term human perspective.

The meaning of "commodity" clearly depends on "relationship with environment" and time preference, but it is also subject to cultural and individual relativity. Even the economic definition of commodity allows for wide variation of likes and dislikes, motivational structure, and so on. A one-hundred-percent-Grade-A-beef superburger may be a commodity to the typical American child but a discommodity for the vegetarian.

Most if not all throughput processes in nature have a common characteristic: they result in the production both of an intended output and an unintended by-product. The unintended by-product in biological metabolism is human or animal waste; in economic metabolism it is air, soil, and water pollution, garbage, noise, and so on. This unintended by-product is one of the major problems associated with throughput.

There are few if any transformation processes which are exempt from this law. A dinner party means immediate satisfaction but piles of dishes and a general mess to be cleaned up later. Moving to another town, organizing the office, even writing a book, all involve products and by-products. For every chapter of the final text of a book there are masses of discarded paper, first drafts and revised versions which litter the wastepaper basket. The curse of the by-product seems to be inherent in the transformation process. The most we can do is to reduce or use the unintended by-products to the fullest possible extent.

The process of economic metabolism varies in form and quality, depending on the mode of production, which in turn depends on the relative intensity in the use of "factors of production"* and the complexity (or lack of complexity) of the actual throughput process.

In a preindustrial mode of production, best typified by the purely agricultural society, there is, strictly speaking, almost no conservation problem. The principal factor of production is land and the auxiliary factors of production are human and animal labor. There is little capital (i.e., machinery) involved. As K. Hayashi has pointed out, there is a built-in brake which prevents agricultural production from upsetting the environment.† First, agricultural production is governed by seasons. Without sophisticated technology and artificial fertilizers, the soil cannot be hurried and will produce once every few months and no more. Second, all waste products in agricultural processes are organic and easily reusable. They enrich the soil. As long as there is crop rotation, and the requisite balance of elements is main-

*Refers to labor, capital, resources, management, and so on.
†"The Industrialization of Agriculture to the Agriculturalization of Industry," a paper presented to the 1975 conference "Limits to Growth," Houston.

tained, the soil can go on producing indefinitely at a controlled rate.

Attempting to interfere with this natural process will eventually invite Malthusian-type equilibrating mechanisms—mainly famine. A glaring example is the 1973 famine in the Sahel region, partially produced by overgrazing and crude attempts at "bullying the soil."

Figure 2 is a representation of the balanced agricultural cycle. All resources used are either renewable (chemical elements in the soil, rain) or practically inexhaustible (energy from the sun's rays). The system is in balance save for the inevitable heat losses due to entropy.

The passage of Western civilization in the eighteenth century from a primarily agricultural (with some handicraft manufacturing) to an increasingly industrial mode of production was, in retrospect, both a blessing and a curse. It was a blessing insofar as it allowed humanity, at least in the West, to have greater control over its destiny, to be free from famine and to shape its own environment. Human beings became increasingly the architects of the earth's surface, transforming, urbanizing, landscaping, controlling. It was also a curse because, in the process (and especially during the second half of the twentieth century), we created a "surrogate world"* which, while giving us a high material standard of living, threatens to engulf and destroy us.

In the words of David Landes:

The heart of the Industrial Revolution was an interrelated succession of technological changes. . . . In this sense it marked a turning point in Man's history. . . . It has been [likened] to Eve's tasting of the fruit of knowledge . . . or the Greek legend of Prometheus. . . . Adam and Eve lost Paradise, but they retained the knowledge. Prometheus was punished . . . for Zeus sent Pandora with a box of evils to compensate for the advantages of fire; but Zeus never took back the fire.

No one can be sure that mankind will survive this painful course,

*First used by E. Goldsmith, editor of *The Ecologist,* this refers to man-made things.

Figure 2:
A PRE-INDUSTRIAL AGRICULTURAL TRANSFORMATION CYCLE

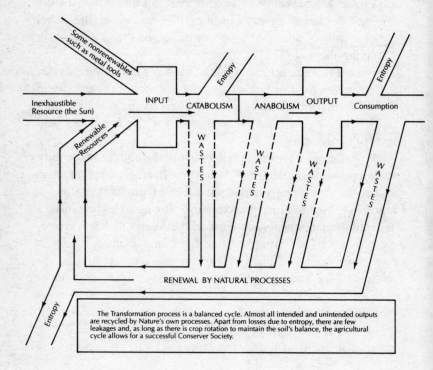

The Transformation process is a balanced cycle. Almost all intended and unintended outputs are recycled by Nature's own processes. Apart from losses due to entropy, there are few leakages and, as long as there is crop rotation to maintain the soil's balance, the agricultural cycle allows for a successful Conserver Society.

especially in an age when man's knowledge of Nature has far outstripped his knowledge of himself. Yet we can be sure that man will take this road and not forsake it; for although he has his fears he also has eternal hope. This, it will be remembered, was the last item in Pandora's box of gifts.*

Landes ends on an optimistic note, but *Limits to Growth* offers a pessimistic view of the ultimate prospects for mankind:

We have shown that in the world model the application of technology to apparent problems of resource depletion on pollution or food shortage has no impact on the essential problem which is exponential growth in a finite and complex system.†

The actual process of industrialization which has characterized Western history for the past two centuries can in essence be described in terms of six predominant characteristics:

1. Unlike traditional agricultural production, industrial production is heavily capital-intensive. It uses machinery in ever-increasing amounts. This culminates in the factory system, in which one gigantic machine, the factory, churns out products, using labor as a mere adjunct to the progressively more automated productive process.
2. Industrial production demands enormous amounts of raw materials. This thirst for raw materials is not dampened by the natural seasons but by the fluctuations of the business cycle, whose amplitude became greater as industrialization progressed in the nineteenth and twentieth centuries.
3. Raw materials in particularly ever-increasing demand are the energy resources. Since the Industrial Revolution was, quintessentially, a substitution of mechanical for human and animal power, the source of this mechanical energy was avidly coveted. When technology dictated steam

*David Landes, *The Unbound Prometheus*. Cambridge University Press, 1969.
†Donella H. Meadows et al., *Limits to Growth* (New York: Universe Books, 1974).

Figure 3
**HAPHAZARD INDUSTRIAL PRODUCTION:
THE BROKEN CHAIN**

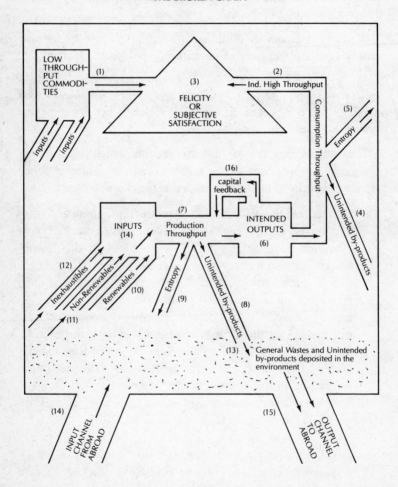

EXPLANATION FOR FIGURE 3
HAPHAZARD INDUSTRIAL PRODUCTION
AS A BROKEN CHAIN

1	Low throughput commodities' channel to satisfaction
3	is the goal of all activity: a subjective state of felicity or satisfaction, represented by a triangle
2	is the channel to satisfaction from high-throughput activities
10,11,12	are input-channels to throughput
7	is the production transformation channel
9,5	are heat-loss channels (entropy)
8,4	are "exhaust channels" where throughput wastes are deposited in the environment
14,15	are input and output channels, to and from the rest of the world. (In Canada, for instance, much of Canadian pollution comes not from Canadian but from U.S. throughput.)
16	is the capital channel. In industrial production a part of the intended output is ploughed back into the throughput process in the form of machinery
13	is the environment with wastes deposited within it

THE BROKEN CHAIN
Considerable waste is deposited in the environment.
Unlike pre-industrial agricultural production there
is no automatic feedback loop closing the chain.
The environmental wastes may accumulate until the
entire system self-destructs.

power, the basic fossil fuel in demand was coal. When technology allowed for internal-combustion engines and electricity, petroleum and its derivatives were necessary. Today's technology is still heavily dependent upon coal and petroleum, the basic fossil fuels.

4. Industry used up great quantities of nonrenewable resources—not just coal and petroleum, but metals— producing the threat of irreversible depletion of these resources.*

5. Industry involves a much longer and more complex throughput than agriculture. This greater complexity increases the production of such unwanted by-products as pollution and garbage.

6. Industry is large scale. The factory system encourages geographical concentration of industrial activity and the growth of cities, accelerating the urbanization process.

In Figure 3, a description of haphazard, uncontrolled industrial production is attempted. The main message of this figure is to show that such uncontrolled industrial production leads to a broken chain with potentially fatal consequences to society. Considerable waste is deposited in the environment, which may accumulate until the entire system self-destructs.

In the figure, the manifest goal of all activity is the attainment of a subjective state of felicity (or satisfaction, high quality of life, happiness) expressed in triangle 3.

Low-throughput satisfaction comes from "enjoyment" rather than "consumption" commodities. A walk through a forest, contemplating a sunset, or listening to birds sing may yield great satisfaction without in the process destroying the forest, the sunset, or the birds. This is in marked contrast to eating a steak, which effectively destroys it, or driving a car, which uses up gas and hastens the eventual depreciation of the car itself. It is high

---

*Fossil fuels are "potentially" renewable, but the renewing process takes millions of years.

throughput. Strictly speaking, all life processes involve some transformation, but the energy expenditure in plucking satisfaction from the song of a bird is, of course, minimal and really insignificant. Flying the Concorde from New York to London for a business appointment, on the other hand, is extremely expensive, in monetary and ecological terms.

In Figure 3, the throughput process itself involves two sub-processes: production transformation (Channel 8) and consumption transformation (Channel 4). Production requires a series of inputs. We distinguish three types: (a) inexhaustibles (inputs in such an enormous supply that they are practically inexhaustible for millions of years. The prime example is, of course, the sun); (b) renewables (inputs which are renewed through nature's own processes if these are not overloaded, such as trees and water); (c) nonrenewables (resources in finite supply which are not renewable within a meaningful time horizon, most notably, of course, the fossil fuels).

Both the production and consumption throughput produce, in addition to the intended output, unintended outputs. First, there are the irrecoverable energy losses due to entropy. Second, there are the general waste products deposited into the environment as if it were a gigantic garbage dump (Channels 5 and 9). The factory produces effluents, but, it must not be forgotten, so does the consumer. Postconsumer waste is even more threatening than some production wastes.

The danger in unrestricted industrial production is its "broken-chain" aspect. Since there is no automatic recycling provision, effluents which escape the concern of any one individual or firm are released, in the throughput process. Everybody is responsible and no one is. The biosphere cannot function indefinitely as a garbage dump; ultimately the system will self-destruct or come to a natural balance through Malthusian means.

Systems which do not possess built-in decelerators or regulators are known as "overshoot-and-collapse systems." The absence of negative-feedback loops leads to overshooting the tolerance limits, culminating in collapse. Thomas Malthus, in the late eighteenth century, fully described these systems. Malthusian mechanisms are the ultimate controllers of "broken chains"; unless the chain is repaired by human hands, nature will do it—and nature's remedies are indifferent to human suffering and pain.

All these considerations lead us to challenge the original formulation of the problem presented in the "limits to growth." We feel this concept is misleading on several counts and prefer the idea of "limits to throughput."

First, if economic growth is to be given any operational meaning it must be tied to the usual measure of performance, the gross national product. But the GNP is a very imperfect measure. To begin with, it measures flow not stock, that is, changes per period of time rather than accumulated production over time. A stock measure would be much more meaningful. A 10 percent growth rate does not mean the same thing for Montreal, already at the limits of congestion, as it does for Ingonish in Nova Scotia. In addition, the GNP measures not only material production but services. An increase in activity by hairdressers, manicurists, poets, and undertakers augments the GNP but does not necessarily raise industrial throughput. Therefore, a lower GNP is *not* a necessary prerequisite for conservation.

Second, even when GNP measures industrial throughput only, it could reflect either conventional transformation processes or recycling processes. The latter could even become major growth industries and raise the GNP, improving the environmental situation instead of worsening it. A recycling channel added to our industrial throughput system is itself a throughput. Here we fight fire with fire. The overabundance of "forward" throughput is neutralized by "reverse" throughput, i.e., recycling.

Because of these facts we feel it is more meaningful to speak of limits to throughput (especially conventional throughput) than of limits to growth. Besides being misleading, the latter expression is discouraging to those who equate zero economic growth with overall stagnation. Limits to throughput, on the other hand, allows scope for selective and conservationist growth.

In *Grow or Die,* an important book on transformation, George T. Lock Land distinguishes three types of growth. The first is additive or accretive, an accumulation of sameness (more of the same). The second is replicative, an accumulation of likeness through reproduction (the prime example is the biological process of mitosis, in which cells reproduce into other identical cells). The third and highest form of growth is mutualistic, embodied in the biological meiosis, in which process all agents change their structure.

Mutualistic growth takes place in social systems when free cross-cultural exchange modifies the mental constructs of all parties. There is no passive or active partner: all change simultaneously. Applied to the interaction of the human species and its environment, mutualistic growth takes place when people change their environment and are changed by it. It is not a one-way process. Technology is one way of doing this and, if applied prudently, can benefit both humanity and the environment. Landscaping, for instance, is modifying the environment but not necessarily for the worse. Therefore, Lock Land argues, it is not growth that is the problem, but the wrong type of growth: accretive when it should be replicative, replicative when it should be mutualistic.

A similar conclusion is reached in *Mankind at the Turning Point,* the second report to the Club of Rome, which argues for organic growth, by which is meant a growth more subtle and balanced than the mere multiplication of shoes, ships, and sealing wax that seems to characterize our high-throughput, mass-consumption economy.

Figure 4

GRAPHICAL REPRESENTATION OF THE LIMITS
TO THROUGHPUT

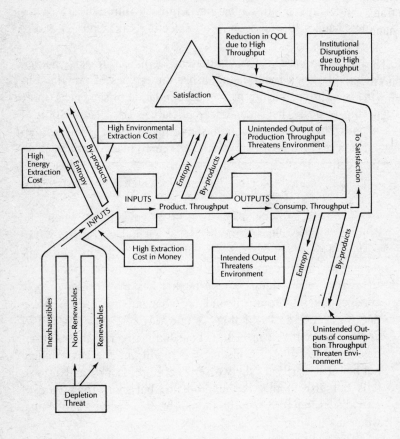

We are in strong sympathy with these views and therefore suggest the reformulation of the basic problem. *The Conserver Society* deals with the limits to a certain type of one-sided industrialization which has characterized our society so far. It will not in any sense propose a total stop in the life processes of social change and intelligent growth.

The limits to throughput can be readily identified; they fall into two major categories: physical and subjective. The physical limits may be further subdivided into input and output limits. The input limits to throughput are established by:

1. *The depletion threat of nonrenewable and renewable resources.* It is intuitively clear that it is possible to deplete nonrenewables, but it is less obvious concerning renewables. However, we can extinguish a certain species of the plant or animal kingdoms by eliminating its conditions of renewal. We can also "overload" a "renewable" water system by polluting it so severely that we destroy the natural mechanisms of self-renewal.
2. *High extraction costs in money terms.* The extraction of raw materials from the soil or the earth's crust is a costly process. The sheer economics of extraction may be an effective limit to further throughput.
3. *High extraction costs in energy terms.* Money cost does not always reflect total cost. In some instances a more meaningful measure is energy cost—the energy cost of extracting a barrel of crude from the Alberta tar sands *vs.* extraction in Saudi Arabia. The energy cost of extraction can sometimes be expressed using the concept of new energy, i.e., gross energy extracted minus the energy used to extract it.*

*A recent weight-reducing diet has advocated the consumption of numerous hard-boiled eggs on the grounds that whereas a hard-boiled egg produces 80 calories it takes

4. *High extraction cost in environmental terms.* Notwithstanding the two measures of extraction cost, there is a further input limit in terms of environmental deterioration. Certain extraction techniques gravely disturb the environment (strip mining, for example). This cost must be accounted for somewhere. When high enough it constitutes a limit to throughput.

The output limits to throughput are as follows:

5. *Intended output threatens the environment.* This situation occurs where geographical concentration of industrial activity, causing high-density urbanization, leads to environmental deterioration. Urban sprawl robs the land of alternative uses and, in addition, may create certain critical chemical imbalances in the atmosphere. The carbon-dioxide/oxygen balance, for instance, may be threatened since it depends on photosynthesis carried out mainly by trees.

6. *Unintended output of production throughput threatens environment.* This category covers the general effluents and by-products of the productive process (scrap, waste, air pollution, noise, poisons).

7. *Unintended output of consumption throughput threatens environment.* Here lies an important limit: postconsumer waste. The consumer destroys the product in the process of consumption and produces garbage, noxious gases, noise, and/or pollution.

The subjective limits to throughput are institutional and value-based.

---

90 calories to digest it. There is therefore a net energy (and therefore weight) loss for the eater. The same idea is present in extraction cost. We sometimes get negative net-energy figures which are not translated into dollar terms. In other words, although a net-energy cost/benefit calculus gives us negative results the dollar cost/benefit shows a profit. Dollar figures are not always meaningful.

8. *Institutional limits to throughput*. These involve functional imperfections of social institutions which are both effects and causes of throughput problems. The inability of the market to price all costs, and the failure of the communication network of society to provide adequate advance warning of dangers, are themselves limits to throughput. In addition, the absorptive capacity of institutions is limited, a circumstance which itself imposes a constraint on further growth in throughput.

9. *Value limits and "other."* In this category lie the value dimensions which, depending on culture and personality, are either favorable or inimical to high throughput. The value system imposes a constraint upon the absorptive capacity of institutions and is a limit to unrestricted industrial transformation.

10. *Human-health limits*. The human body, like most biological systems, maintains homeostasis by keeping within fairly narrow limits such life indicators as heartbeat, brain waves, and blood pressure. Throughput may have positive or negative effects on these indicators; it can cause emphysema or create life-saving synthetic vitamins. What should be kept in mind is that throughput, as a process which is intended to provide for human needs, must be of the type and quantity which in fact serves and does not frustrate that purpose.

In a sense, the limits to throughput are not unlike the limits to eating. In the latter there are input limits based on the availability of food, its extraction cost, and its processing cost. There are output limits which include both intended and unintended outputs. Capacity to consume governs both input and output limits, spaghetti-eating contests notwithstanding. Finally, there are also subjective limits to eating, which is both a biological and a cultural experience. We eat certain foods, delicately seasoned, accompanied by appropriate drinks and in a pleasant setting. These cultural dimensions impose limits to the quantity and quality of what we can eat. *Mutatis mutandis,* the same is true for economic throughput.

We can restate the basic problem for which the conserver society is a possible solution: there are effective limits to one-way throughput that must be respected if the system is going to survive. If these are not respected, then Malthusian mechanisms will restore the balance the hard way—through social disruption, famine, asphyxiation, disease, or other natural disasters.

Let us now consider, in greater detail, the physical and institutional constraints to throughput which North America is likely to face in the coming decades.

Sammy Squander Contemplates the Big Rock Candy Mountain

Ever since that day at the psychiatrist's the Big Rock Candy Mountain had been on Sammy's mind—and with distressing frequency. He had begun to dream about the endless climb, his feet seeming heavier and clumsier as he stumbled among the objects of his "up-and-coming" happiness. But even if the way ahead (up the mountain) was a glorious vision, still it was terrifying to see all his future possessions in one overwhelming glut. The feeling was reminiscent of the many times when, as a child, he had eaten too much candy—as much as his mother would give him (which, when you considered it, was as much as he would eat). A new fear, an uneasiness, was added to the habitual feelings of dissatisfaction he had suffered before. And a further difficulty . . .

As with all mountain climbing, especially if it's uphill all the way, there is a temptation to look over one's shoulder, to glance back down. Sammy thought he knew what he would see if he did look back—all the things he had ever possessed. He could imagine the hundreds of pairs of shoes, the sports equipment, games, toys, at least six television sets, countless pairs of jeans, stereos, records, books, ten bicycles, literally hundreds of toy cars, and quite a few full-size ones. He had already had a glimpse of the accumulation when he moved out of his parents' place to his new one-bedroom pad, a few years before.

That was one of his immediate problems. It had really been just a prestige trip to take a bachelor apartment in one of the new airproof buildings—and it had brought precious little prestige. Sammy couldn't even invite his friends over. What with the eight speakers on his new octophonic sound system, the decorator lamps, the thirty-inch color TV, the $1,000 water bed, skis, clothes for every event, the newest in hanging plants, he could scarcely move in the place himself. The real worry, of course, was the expense.

When even the indulgence of his doting parents had seemed restrictive, and Sammy had moved out to his own pad, he had been caught in an upward spiral of living costs. His life-style had drawn him into the expensive restaurant-discothèque scene, where he paid high prices for substandard fare—so much so that a full range of vitamin supplements had become a regular and necessary expense.

Sammy spent so much time trying to assemble the best and latest sports equipment that he had almost no time to use it. Just this last season he had taken about a week to find the right ski boots—and had spent only one afternoon on the slopes. He knew that he was getting badly out of shape but couldn't seem to find time to exercise. Of course, Sammy never walked anywhere. After all, he had just invested in the new Super Stratobird Special 2 + 2 with moon roof, power everything, 7½-liter engine, and a four-barrel carburetor. Like many other things in Sammy's life the Stratobird was a disappointment—in performance and prestige. A gas guzzler, it was nevertheless a loser—430 cubic inches of engine but no acceleration. Emission controls, and an inefficient carburetor system, power windows, power brakes, power steering robbed the monster of any pep. Unfortunately, Sammy couldn't even use Stratobird 2 + 2 for double-dating as he had intended. In spite of its impressive exterior size, the car was cramped inside, especially for the rear passengers, some of whom had called it "2 in comfort, 2 in agony." Even that great macho kick Sammy had felt when people admired the car had lasted only its first week. It was the same with almost everything he bought.

Sammy and the people around him—who weren't even his friends—were involved in the uphill struggle of trying to impress one another. And none of them were going anywhere. There was really no "top" to be reached—just a merry-go-round of mutual envy, mutual oneupmanship and an empty life of communication-in-clichés. Worst of all, perhaps, Sammy was finding it difficult to communicate with his parents and old friends. With every new debt incurred the atmosphere in the Squander household was getting colder and colder. When Sammy visited now and again (for a

good meal, as much as anything else), he always came away won-
dering if the alienation was caused by his anger at his father's increasingly tight-fisted attitude or by his own embarrassment and
even guilt.

What with the unpaid bills piling up, his parents' disapproval
and the recurring dreams about the Big Rock Candy Mountain
(which was becoming more like a nightmare) Sammy was beginning to question his whole life-style. For the first time, Sammy
even began to have doubts about Dr. Fraudoong. He had often
heard that psychiatrists were no use but had just assumed that this
was said by people who couldn't afford the best. Now he began to
wonder. His habitual reliance on the idea that anything that cost
money must be worth getting was gradually breaking down. The
vague, slightly garbled memory from economics classes, that real
value was reflected in price, began to seem confusing. "Does
something cost a lot because I need it desperately," he asked
himself, "or, do I think it can help me because it costs a lot?"
Sammy was perplexed. He thought he'd talk to Fiona about it
before he helped her make her summer vacation plans. The trouble was that she wanted to visit half a dozen exotic places and
Sammy couldn't afford even one of them. He had resigned himself
to spending his summer alone in the hot, humid city while trying
to pay off some of his debts.

When they talked later—over something Fiona had concocted
with Caviar Helper—he had trouble getting her to understand his
problem—she just offered to lend him money. They finally
dropped the subject and got to talking about their old friends,
discussing their various summer plans. On the way home from
Fiona's Sammy reflected that his situation would be all the more
bitter when contrasted with those of his friends, but he was trying
to take a constructive attitude. He thought, "They must be doing
something right. Maybe I can learn from people like Angus McThrift,
Lenny Lease, Mister Middleton, Rita Righteous, and Solo Seleccione."

He had no doubt he'd receive hastily written letters and brilliant
postcards from them all. They knew about some of his problems

and would, as always, be offering advice. Short of tripping out on drugs, adopting a new religion, going in for advanced yoga, or consulting a spirit guide... "Now, *that's* an idea," Sammy thought.

4. THE PHYSICAL LIMITS TO THROUGHPUT

Technology, in the broadest sense, is the means by which humanity interacts with the environment. It consists not only of the physical "hardware"—machinery, computers, and so on—but also of the ideas and inventiveness of the human mind which give rise to the hardware. Indeed, if we can consider humanity as separate from the environment (just for the moment) it is possible to view the three elements in the interaction as in continuity: technology, being made up, on the one side, of ideas and, on the other, of physical materials deriving from the environment, straddles the division between the human and environmental elements.

Given our anthropocentric bias, technology (at least in the West) tends to be regarded as the means by which people *act on* the physical universe to convert its energy and materials for human use. It is also true, however, that the physical environment is an active partner in this process. In a cold climate, for example, the environment dictates not only the need for shelter but also the resources and technology available to build it as well as the repercussions on a society of the use of that technology. Indeed, since human beings are extremely sensitive biological systems which can be maintained only within a narrow range of temperature and only if supplied with certain amounts of specific nutrients, there is a relatively small land area on which they can survive without technology.

Throughput, which is essentially the conversion of energy and materials into different forms, is going on continuously, not only by human intention but spontaneously (as far as we know) in the physical environment. Even in the most arid desert there is alternate cooling and warming of the land by night and day; there are organisms alive in the sand; geological change is continuous. Nature's throughput, however, whether by some grand and intricate design or by spontaneous and successive weeding-out processes, is extremely well organized. It operates at a level of sophistication and complexity to which the human species cannot even aspire. The most forbidding physical limit to man-adapted throughput, indeed, is not the palpable and offensive pollution which occurs both temporally and spatially close to industrial plants: we have immediately within our grasp the effluent controls to deal with that. A much more serious problem is our lack of understanding of the extensive, long-term effects on the atmosphere, not only of the initial pollution but of the attempts to eradicate it. It is only recently that, beginning with Rachel Carson's *Silent Spring,* an insight into this lack of understanding has been gained. Perhaps the least-appreciated limit to human adaption of natural processes lies in the human brain.

There is a tendency in assessing the sophistication of our technological systems to use a time dimension for purposes of comparison. If the highly complex systems of the second half of the twentieth century are compared with the systems of primitive times or even those of the early stages of the Industrial Revolution, the new technologies seem immensely sophisticated. The comparison, however, is almost irrelevant. The only relevant comparison is with the biological and ecological systems with which technological systems must interact. The latter, viewed in this light, are woefully inadequate; they are light years behind the complexity of living systems. To take such a negative attitude, of course, runs counter to the view of man as a God-like creature, and intentionally so, but it also allows for tremendous scope within that limit. Reason is part of humanity; humanity is part of nature. The energetic application of intelligence (which is better than renewable, since it improves with use) to the problem of survival results in technology. Nothing, therefore, could be more natural than technology, except perhaps the application of more intelligence to its use.

On the human side of the humanity-technology-environment interaction, there are thus both physiological and psychological limits to throughput. There are additional psychological, moral, and cultural constraints which do or should operate when human individuals are grouped in societies.

Those constraints which stem from the nature of a society are most generally and directly manifested as economic limits to throughput. Particularly important are relative prices to the consumer, and the cost of initial capital investment, the second greatly influencing the first. Relative prices of alternative resources may change, literally, overnight. This is exactly what happened in 1973, with the important increases in the price of petroleum making hitherto unprofitable alternative sources of energy suddenly profitable. That part of the economic limit to energy conversion is thus highly volatile and subject to change. Fluctuation in prices, therefore, influences the kind and quantity of throughput and can do so fairly suddenly. However, the changeover to a new type of energy conversion, or even the further development of an existing type, invariably involves high initial capital investment. The first stages of a general transition from fossil fuels to nuclear power, for example, would perhaps be prohibitively costly, thus setting at least a short-term limit on freedom of choice.

The limits just described reflect the nature of a society and emanate through technology toward the environment. Their importance must not be underestimated since there is considerable variation in the approaches made by different migrant peoples to the same ecological situation. A striking contrast exists, for example, between the manner of living of the ancestors of North American Indians and Eskimos and that of the mass-consumption society created by the relatively new wave of European peoples. In the former, the human society accepted (or was forced to accept) a kind of accommodation with nature which kept the population small and the culture appropriate; in the latter, the human society's will (to survive?) has been imposed, through technological intervention, on the physical environment. Clearly there is more than one way to live in any given ecological setting.

The purpose of this chapter is to isolate and examine the manageable, purely physical variables and to indicate the constraints which they place, and the scope which they offer, to possible human behaviors in the physical environment. We must not, however, overestimate the significance of physical variables simply because we can estimate their quantities. The geologist and the engineer may measure for us the exact position, supply, and amount of energy required to extract and transport a certain mineral. The ecologist may tell us the cost of cleaning up the environmental effects of these actions. These measures appear to give an objective and unchanging "value" to the mineral; but that, in fact, can be determined only by the society's value system. In other words, the value depends on the perceived need for the mineral. If we were to indicate, for instance, how much bituminous coal North America possesses we would not be automatically suggesting that it should or should not be used. This, we contend, depends on the human variables—the ideals, preferences, and attitudes within the society or societies which have access to the bituminous coal. The hard facts themselves, however, can perhaps be effective in appropriately modifying the beliefs, and therefore ideals and even preferences, which are brought to bear on such decisions.

Now, which are the most important hard facts? A myriad of statistics, all of them relevant in some degree, appear in reports produced monthly, quarterly, annually. To justify a choice among them we must make our reasonable assumptions explicit. Even if we do not continue with the status quo for the next fifteen years, we will probably still use the same type of energy and materials involved in current throughput—the same kinds of housing, transportation, industry, entertainment. We may quickly learn (as President Carter suggested in 1977) to use less fossil fuel and to use it more efficiently, but, for a few years at least, we will continue to use it. Equally, we can assume that alternative sources of energy, at present in the early stages of technological innovation, will be introduced, but we cannot rely on radical and instantaneous advances.

Just as we cannot assume spectacular technological breakthroughs, neither can we take into account the possibility of unforeseeable natural catastrophes such as sudden climatic change. To express it more directly, it seems reasonable to suppose that the human brain and the ecological setting will continue as in the recent past and at present for at least the next fifteen years.

Perhaps the outermost limit which the physical environment places on energy and materials conversion was first expressed as the second law of thermodynamics. This so-far-unrefuted law has been generalized as the law of entropy to apply to all energy conversions and transfers. Stated simply, it means that in all processes in which energy is changed (in its location or its form) some of the energy is dissipated. It is not "lost" in the sense that it disappears from the universe, but rather it becomes effectively unavailable, usually because it is diffused and therefore impossible (or very difficult) to harness. The low-grade heat, for example, created by the friction of shoe sole on sidewalk is not destroyed; it simply exists in such low concentration that it is useless. Entropy seems to be nature's way of charging for the "service" of energy conversion. It places an absolute limit on throughput, since it means that we can use in effect something less than 100 percent of the resources which seem to be available. Some small percentage must be paid for the energy-conversion process, even that carried out by the human body. There is some evidence to suggest that man-made technological devices are more entropic, that is, wasteful, than biological and ecological systems, but nature is not entirely exempt from its own service charge. Technological systems vary in the amount of energy loss they cause, and it is for this reason that entropy could also conceivably become a limiting factor on the types of throughput from which a society would choose to produce the energy it requires. We do not yet have sufficiently precise measuring devices to divide the inefficiency of an industrial process between entropy and non-entropy loss but, given producers' own interest in reducing waste, it is highly probable that there is a correspondence between entropy loss and total energy loss in an energy-production process. We know, for example, that of the conventional power sources coal is the most efficient, with only a 5 percent energy loss, contrasting dramatically with tar-sands petroleum, which loses 23 percent.

The law of conservation of energy states that energy cannot be destroyed. Equally, it reminds us that energy and matter cannot be created. There is an upper and absolute limit to throughput inherent in the nature of the earth or even of the universe. We should remember that the activity commonly spoken of as "production" is in fact extraction, conversion, transformation, or some other noncreative process.

The average American (or Canadian) consumes the energy equivalent of more than 50 barrels of petroleum every year (twice that of the average European). Energy used per capita is divided as follows: residential, 16 percent; business and administration, 12 percent; industry, 23 percent; transportation, 20 percent; energy industry, 25 percent; non-energy utilization of energy resources, 4 percent.

There are at least two startling figures in this breakdown.

Roughly one-quarter of the total energy "produced" in the United States and Canada is used in the further production of energy. The producers are their own best customers. It means, effectively, that to extract, refine, and transport three barrels of oil to the retail outlet the equivalent of one barrel of oil must be sacrificed. This cannibalistic situation will persist as long as the energy industry remains, like most other businesses, energy (rather than labor) intensive or until antiwaste measures are taken. The other startling figure is transportation. One-fifth of total energy used in Canada and the United States is expended on moving people and things from one location to another. It has been suggested recently that the figure should be much higher, perhaps even twice as high, since it does not include the energy consumption of transportation-related industries such as garage building and parking-lot services. Three-quarters of the energy expended on transportation provides fuel for highway vehicles—mostly private cars. There are more than two cars for every five persons in the United States. Since two cars are designed to hold at least eight people, there seems something sadly wrong.

Residential energy demand is high for two reasons. First, heating requirements are necessarily very high in Canada and in the northern parts of the United States, with their severe winters. An even greater drain, however, is caused by the proliferation of electrically powered domestic appliances. Convenience, comfort, entertainment, the avoidance of use of human energy—all are characteristic of the mass-consumption society. A heavy price is paid in nonhuman energy expenditure.

Within the industrial sector the heavy energy consumers are, in descending order: (a) pulp and paper (newspapers, tissues, shopping bags); (b) primary metals (car bodies, laundry appliances, etc.); (c) nonmetallic products (plastic bags, shoes, sealing wax, etc.); and (d) the chemical industry (including pharmaceuticals).

It is generally agreed by the optimists and by the prophets of doom alike that if present trends continue the critical period of energy shortage is from now until around 1995 or the end of the century. With domestic supplies of petroleum and natural gas running dangerously low, the immediate problem is to bridge the gap with alternative sources and/or to reduce the gap by conservation measures or by increasing dependence on foreign sources. Since it is not possible to predict exactly the supply to be derived from currently known alternative sources, which require new technologies, and since importation must be regarded, at best, as uncertain, it is sensible to conserve. This is especially so since, if we begin now, a reduction of 10–30 percent in energy expenditure can be progressively brought about simply by the elimination of waste, *with minimal interference with our present way of life*. (More detailed discussion of waste and its avoidance can be found in later chapters.) Alternative sources do offer some scope and must be quickly but carefully developed, since it is largely on these that we will depend by the beginning of the twenty-first century. Their exploitation will not be problem-free.

Coal, although it is in abundant supply, carries high costs in both environmental degradation and transportation.

Hydroelectricity, although it is a clean and renewable form of energy, involves high environmental and initial investment costs.

Geothermal and tidal power can make substantial but mainly regional contributions.

As far as nuclear power is concerned, advantages such as assured and abundant supply, competitiveness with other energy sources, and diversity of application must be set against the worries inspired by its potential ill-effects. No other source of energy, today, offers such promise of immediate and rapid development, since there is no assurance of increase in hydraulic energy or fuel from hydrocarbons. A full assessment of all the costs and all the benefits is yet to be made.

As a source of energy, solar power, which is diffuse and intermittently available, does not yet lend itself well to exploitation on a large scale. There is some chance that it could break into the commercial and residential heating market, although the design of buildings in situ and the conservatism of the construction industry make it seem unlikely that this could occur on a grand scale in the near future. It is, however, on solar power with all its derivatives that hope for a decentralized type of energy rests: it is the best long-range solution to the energy problem.

This is a thumbnail sketch of the immediate outlook for energy. To understand the environmental limitations which have, in our particular time and place, given us these options it is helpful to place them in the more general framework of renewability and accessibility of resources, pollution, and population.

Renewability

In North America as in most of the industrialized world there is heavy demand for a group of substances which, for human intents and purposes, are nonrenewable. The designation is not a particularly good one, because to suggest that a substance cannot be renewed is to attempt to contradict the laws of indestructibility of matter and the conservation of energy, and because "nonrenewable" is not exactly what we mean. What we really mean is that these substances are not in continuous supply at a rate which is of significance to the perception of human beings. Solar power is seen as a renewable resource because we notice the sun's dependable reappearance day after day. Fossil fuels, on the other hand, are thought of as nonrenewable, partly because their natural rate of accretion is agonizingly slow on a human time scale, and also because they are used in such a way as to radically change their form. Not only is there a limited stock of petroleum available under the earth's surface at any given time, but when it is burned in automobiles it becomes virtually irrecoverable. The heat derived from burning it is transformed radically into mechanical energy to move the vehicle, and particles are deposited in the atmosphere as pollution. We do not yet have a working technology which can capture and recycle the wastes. The common metals are also thought of as nonrenewable, although they are not generally used in such a way as to be entirely irrecoverable. Obviously the metal in a car body, although its shape is designed for a specific use and it has been adulterated by paints and other materials, is not "destroyed" in the same sense as the petroleum is. It is conceivable that a car could be recycled to become part of a washing machine, a bicycle, or a new automobile. The same metal, of course, *can* be used in such a way that it is irretrievably dissipated, as with a file, which simply wears down, its particles being dispersed about the atmosphere.

The renewability or nonrenewability of a resource, therefore, depends on the rate of use relative to the rate of replenishment by nature and also on the manner of use. In *Small Is Beautiful,* E. F. Schumacher has suggested an illuminating distinction between "capital fuels," the nonrenewables, and "income fuels," which are renewable. When a nation such as the United States depends for more than three-quarters of its energy needs on burning fossil fuels, the well-known folly of living off one's capital is being committed. Even if the fuel is imported in an apparently steady stream it is still "capital," however "foreign" it might be.

The squandering of limited supplies is not, however, the most ironic aspect of the depletion of nonrenewables. Fossil fuels and the simple minerals are highly versatile. They can be combined with many other substances under a variety of conditions and thus can be adapted technologically for many exciting new applications. When petroleum, for example, can be used in pharmaceuticals, plastic products, even as food, it seems almost primitive to annihilate the whole supply in the simplest and most destructive way—by burning it.

Accessibility

It is very difficult to define the extent to which throughput is constrained by the limited nature of the supply of nonrenewables in the physical environment. There are large areas of land and sea not yet explored; there are apparent reserves untested for their potential; there are even "actual" reserves whose quantities are still subject to debate. Anyone in the mining industry will admit that reserves are not "proven" until they are "on the shelf." There is, in short, a great deal of uncertainty. "Total" reserves of any particular mineral would seem at first glance to be the only significant figure if a threat of depletion is present but, in fact, even well-educated guesses differ by multiples as high as 200,000. The *Limits to Growth* report assumed that existing estimates of mineral reserves understated the "total" figure by a factor of 5; the Hudson Institute assumed the understatement to be by a factor of 1,000,000. Given such wide diversity of opinion, it is fortunate indeed that the total figure is in fact of very little significance in itself. The actual amount, in tons, barrels, or whatever, is meaningless without companion figures describing what we will call "accessibility."

There are at least three elements implicated in the accessibility—to human beings—of a resource (nonrenewable or otherwise).

1. *Concentration.* Mineral deposits vary greatly in their purity. High-grade reserves can be taken out of the ground almost in their pure state, unadulterated by admixtures. Low-grade reserves are so mixed with other materials that they represent only a small proportion of the matter which is "found." It is easy to see that to "produce" the pure mineral is, in each instance, a very different problem. High-grade sources may require only a washing procedure to clean off impurities, but the extraction of low-grade reserves requires high energy input, usually of fuels threatened themselves by depletion; massive waste materials must be

disposed of, the wastes themselves often posing a threat to the physical environment and to neighboring human communities.

2. *Depth*. A mineral may be of high-grade concentration but lie so deep under the earth's surface that extraction must overcome the same type of technological, energy-depletion, and environmental difficulties as are associated with low-grade deposits. Waste disposal might be somewhat less of a problem since the intervening layers can be tunneled through, but this, at least with conventional methods, requires tremendous force and uses up considerable amounts of energy reserves. Also, the tunneled-out material is often neither replaced nor put to any use, becoming just another polluting eyesore.

3. *Distance*. The transportation of a substance from point of extraction to point of use involves an energy cost (usually translated into money terms and added to the user's price) not only in actually moving the goods but also in the creation of pipelines, roads, tankers, transport trucks, and so on. Increasingly, new projects are being built in areas of severe climate far from urban centers where the energy and minerals once transported are finally used.

Concentration, depth, and distance collectively define the accessibility of a resource. A given, finite amount of oil, for example, could lie close to the surface, in a pure state, in a location not far from point of use; on the other hand, the same finite amount of oil could exist in low-grade deposit, deep below the surface, and far from where it is to be used. The energy value of the latter deposit is the same as that of the first, but we have to subtract the cost associated with extraction, cleaning up pollution, and transportation. Full-cost pricing, which will be described later in this book, can translate the real energy costs into economic costs to be charged to the user. It cannot, however, charge for the social and health costs involved in extraction of relatively inaccessible resources. It could not, for example, take account of the deterioration of workers' health from pollution associated with the mining of low-grade resources (unless one would cynically point to workmen's compensation), or of the loss of cultural continuity involved in the displacement of remote communities, or of the incalculable environmental effects of extensive interference with fauna and flora sometimes occasioned by monster projects.

Even if the quality-of-life dimension is ignored, the "energy sense" of some major projects recently undertaken is highly questionable. The total efficiency of an operation all the way around the throughput cycle, from the construction of the well head (in the example of petroleum) to the operation of the car engine, even to the environmental effects, must be considered and not simply the monetary profitability from well head to gas tank.

Inaccessibility, therefore, should be considered along with nonrenewability as a throughput limit which stems from the environment. Renewability (and its opposite) is governed by rate of use and replenishment and therefore depends on the dimension of time. Accessibility (and its opposite) depends on location and concentration and therefore is governed by the dimension of space. The philosophical implications of regarding resources in this way are intricate, deep, various, and probably best abandoned for the time being. The salient points, in this context, are that renewability should not be thought of as an inherent quality of certain substances and that accessibility must also be taken into account.

In discussing accessibility we have tended to use examples of substances which are called nonrenewable—oil, coal, minerals. This does not imply that only nonrenewable resources are ever difficult to get at. Renewable resources, notably solar, tidal, wind and hydropower, can also be highly inaccessible. Indeed, most of these are so diffuse that they may be said to be in low concentration, making harnessing for intensive use almost impossible. They may also be found—and plentifully—in areas remote from urban centers where the "need" or desire for them exists, thus presenting the distance problem.

However, the observation that some renewable resources are diffuse and remote does not alter the fact that a great deal of this kind of power is readily available, in high enough concentration (for some purposes) precisely where it is required. To say, as many do, that the era of cheap energy is past is simply to admit that we have used up all of the energy that is highly concentrated and handy. Along with, perhaps instead of, the furious headlong rush to exploit the remoter highly-concentrated sources, maybe we should also be investigating the adjustment of our industrial and urban patterns to take advantage of the renewable resources that have been available all along but have been virtually ignored in recent centuries. The soft technologies recommended by E. F. Schumacher may be "appropriate" not only to the human user but also, in the contemporary context, to the kind of energy available. Diffuse, low-concentration sources of power, by their very nature, recommend small production and consumption units widely distributed in the physical environment.

It seems also to be true that pollution problems can largely be avoided by the adoption of a decentralized system of small-scale, but nevertheless highly sophisticated, technological devices. It is the *concentration* of pollutants around gigantic urban and industrial complexes which seriously threatens the physical environment. Indeed, the pollution problem shares with the resource-depletion problem the two-dimensional constraint of time and space operating together. It is the rate (so many tons per day) at which we emit pollutants into the atmosphere and the dumping day after day, in the same place, which overload the recovery powers of air and water.

Ironically enough, the determination to exploit and thus deplete virtually inaccessible nonrenewable resources creates corresponding pollution, which itself adversely affects precisely those renewable resources on which we depend for our very survival and on which we could rely for power.

Pollution

To give some idea of the complexity of the problem of pollution, a brief rundown of complicating factors follows.

A stream can take only so much dumping per day. Its flow, and therefore the extent to which its flow can be impeded without being stopped altogether, is limited.

Some pollutants are absorbed or eaten by organisms but, since these, like everything else in nature, are limited, so also is their capacity to neutralize pollution. Magnesium, for example, is dumped into a river, is eaten by a fish, which is eaten by a person, who dies of magnesium poisoning. Such deaths have already occurred in parts of Japan and Canada from exactly that sequence of events.

Some pollutants simply have no natural enemies existing in the body of water, air, or land into which they are dumped.

Some substances are harmless enough in themselves but become dangerous when combined with other substances. This means that it is difficult to assign responsibility, for example, between two chemical companies dumping in the same lake.

Still other foreign substances which are potentially polluting become harmless when intentionally neutralized.

It is tempting to divide pollution into that of air, fresh water, sea, and soil, but as Rachel Carson has so well documented, pollution moves not only within or through bodies of land, water, and air but also from one to the other. Pollutants in water and even in the soil evaporate into the air to be carried by the wind to distant points on the earth and into the atmosphere.

Pollution in general is no respecter of boundaries, thus creating extraordinarily complex intergovernmental and foreign-relations problems.

Noise pollution, even from relatively innocuous throughput like road building, is disturbing and even dangerous.

It is feared by some experts that pollution from the burning of hydrocarbons can actually bring about (or distort natural) climatic change, even causing drought.

Pollution is sometimes mainly an aesthetic problem, more or less intolerable, depending on the perceptions of the local people.

The most dangerous pollution of all, radioactivity, potentially comes from our most abundant power source, nuclear reactors. It is probably true to say that the deeper the technological intervention, the more hazardous are the unintended by-products. At least, it is dangerous for the by-products to be regarded as unintended (or, in other words, to be disregarded). If it is decided that having lights on in empty rooms is important enough so that future generations must risk debilitating disease, genetic defects, uncontrollable terrorism, then radioactive pollutants are being regarded as merely inconvenient.

The way in which unintended by-products are dispersed about the physical environment is as complex as the workings of the physical environment itself. "Why, then," you might say, "a little pollution is a natural part of the environment and the system should be able to cope with it!" This is true, up to a point—and rests on the surely valid assumption that if carelessness is part of human nature its results are a natural part of the physical environment. However, we would suggest that just as all the other facets of the situation we have outlined are limited, so also is the extent to which the ecological system is stable under dumping of unintended by-products. The environment can put up with only so much stupidity without being altered fundamentally.

The ecological system may indeed cope with massive pollution, but how? We must abandon the notion that humanity is a favored species and will continue to be provided with vital life support in any new arrangements the ecological system makes. Even if we believe in a manlike God, a Supreme Being, a Prime Mover, there seems no particular reason for him or her to perpetuate a species which unnaturally neglects to use its God-given intelligence. The clue to the stupidity of overpollution is in the designation "unintended." The by-products are unintended only in so far and for as long as we do not know they are produced. As soon as we become aware of them they are *intended* by-products; once having discovered that any particular industrial process gives off an effluent that inevitably destroys a species population, if we continue to use that process we *intend* to destroy that species, even if it seems less consequential than the primary objective of the industry. The species destruction has come out of the unfathomable realms of chance and into the clearly delineated realms of choice.

Population

As discussed earlier, human beings are delicate biological systems that live only within narrow limits (in the physical sense). There is only a very limited land space upon which we can survive, and then only in a very simple fashion, unless we are interlocked with technological systems which are, in turn, interlocked with ecological systems.

Of course, there is an outer limit which the total system acting on human beings has placed on human population. The human male can produce only so much sperm; the human female can bear only so many children. This does not appear to be a limit that now or in the near future we will strain against. High-population ambitions seem to belong to earlier centuries. Other limiting factors are the scarcity of materials and energy to support human life and the occurrence of disasters, predators, etc., which threaten it. Given the environmental limits to throughput, it would appear that they are also the relevant bounds to human population. The latter, however, should not be understood merely in terms of numbers—obviously the throughput per person is equally relevant here. These variables are sometimes referred to as the multiplicand, number of persons, and the multiplier, per-capita throughput. To put it more graphically, population can be seen as a rectangular piece of two-way-stretch fabric. The more it is extended in one direction the less it can be pulled in the other. The overall maximum area is a constant.

Needless to say, if we do not accept the premise that the physical universe itself is limited we will not accept this picture of the population problem, no matter how graphically it is drawn. If, however, we do assume that the substances on which survival of human life depends are limited, human life is limited. Let us consider briefly the ways in which it can be constrained. Conception can be prevented by birth control, an infertility imposed artificially by individuals or brought about by natural processes. In the first instance, it is interesting to note that those who choose not to have children at all "select out" their own descendants from future generations in a haphazard fashion.

Life is often shortened by Malthusian mechanisms—war, disease, famine—by natural disaster, such as earthquake, volcano, flooding, by group or individual human actions, such as genocide, homicide, suicide, or euthanasia. Population, of course, can be "designed" either deliberately or inadvertently. Disease prevention and birth control together make for an aging population. Tests during pregnancy, for genetic defects and/or sex, with subsequent abortion, can distort the population. Artificial insemination, genetic counseling, even genetic engineering, cloning, could all conceivably place limits on the composition of a population and therefore on its size. In a perhaps more innocuous vein, population can be contained, even in areas where disaster threatens, by economic and institutional means, such as by immigration laws.

It is indeed the unevenness of the world's population and resources distribution that makes consideration of it so complex. We cannot assume that there is an optimum international standard of resources per capita—simply because both the resources and the values vary so greatly from one country to another. The metaphorical two-way-stretch fabric, if representing the entire planet's human population, is not of even elasticity throughout— its flexibility is not homogeneous. There is a very obvious but apparently not-too-pressing moral question concerning the distribution of basic resources between the rich countries with relatively small populations and the poorer countries where the world's inhabitants are heavily concentrated.

Even if we focus on the North American continent, which is by any standards not the most beleaguered area, the population problem is extremely complex; there are myriad questions to be faced and many different ways of viewing them.

Should we plan ahead to provide for future populations based on growth rates extrapolated from past and current trends?

If so, for which material standard of living?

Should we measure probable future availability of resources to maintain our present way of life and thus decide on an "appropriate" population?

If we decide on an appropriate population should we restrain natural growth and/or immigration in order to achieve it?

These and many related questions are unanswerable until we deal with the implicit value questions underlying them. The problem of population, clearly, is fraught with psychological, cultural, and ethical intricacies, and consideration of these must precede any decision on policy.

"Population" straddles the humanity-technology-environment spectrum and must be approached, so to speak, from both ends. It is constrained by values in the human society or societies concerned; it is also constrained by the physical limits to throughput. Population is central in the production-consumption continuum, since people may be regarded not only as human resources but also ′as the consumers of other resources. Children, generally speaking, are a commodity to their parents; as the route to immortality and as proof of ''manhood'' or ''womanhood'' they are a direct source of ego satisfaction. Traditionally, especially in societies with labor-intensive technologies, children have been regarded as a source of wealth and security; wars have been won and nations glorified simply by superior numbers of dead and alive young men. Children are also consumers, and are such to an extent dictated by their parents' value system and also by the availability of resources. It has been estimated, for example, that a Canadian or North American child consumes fourteen times as much energy and creates fourteen times as much garbage as a child in the People's Republic of China. There is little to suggest that one is happier than the other.

The greatest happiness for the greatest number could, it seems, be achieved by devising a measure of quality of life from the accumulated wisdom of nutritionists, medical researchers, psychologists, and sociologists. This could designate, at least for the material aspects, the optimum quantities and qualities of the various commodities required for the individual human life. How much food? Of what type? How much space? How much warmth? How much air? How clean should the air be? How much fresh water?

We could then define the outer limit to population, determined by environmental factors: the population of a nation at any future date should not exceed the number which can be provided for by the projected available supply of the scarcest of these commodities. A tougher constraint, at least in the North Atlantic world, would be the values boundary. To the extent that beliefs, ideals, and preferences in a society allow the freedom to use vital life systems for nonessential purposes, the limit to population must be foreshortened. If fresh water, for example, were this hypothetical scarcest of substances, as it has recently been in some parts of the world, that resource's supply could not merely be divided by the recommended per-capita consumption to determine an appropriate population. Obviously water used for watering lawns and washing cars is no longer available for drinking. Spring water polluted with poisonous chemicals is unsafe. A population goal must be modified to take into account the probable continuance of present behavioral trends. Of course, institutional constraints (which will be discussed in later chapters) can be (and currently are being) used to change those trends, but that in itself depends on change in the institutions themselves, on the leadership which directs them, and therefore, finally, on the value systems of the society itself. In short, even if we measure precisely the environmental limits to throughput we cannot ask "How many people?" without first answering "What kind of people?"

We have attempted to indicate the physical limits to throughput and to illuminate their nature—"absolute," as determined by apparently universal laws, or "flexible." The empirical facts, or rather estimates of them (so many BTUs, barrels, tons), are largely ignored here since they are appropriate only to studies (usually operational) concerning the availability of particular resources. It is only by detailed explanation of the conditions surrounding that availability—its concentration, depth, and distance—that the "amounts" are meaningful. Our main concern here is to point out that there are physical limits to throughput of any given resource, to indicate these bounds, and thus to ensure that the important variables which may meet, collide, and clash with them are taken into account and calculated, as far as possible, when such studies of availability are carried out. The absolute physical limits, actual amount of basic resources and entropy loss, are ineluctable. No matter how well we play alchemist and change this (in plentiful supply) into that (in short supply), we will pay the energy price of the change. Of course, this uncovers our implicit assumption that the universe itself is finite rather than infinite. (The justification for this supposition is that if we take the other, to be wrong is dangerous.)

The flexible limits could be called the technological limits to throughput, to renewability and accessibility, to environmental pollution, and to population size. Each of these has a range within which it can change according to human behavior. They are interrelated, interdependent variables which collectively enclose a "space of operations" within which choice can be exercised by a human society. As we increase the strains in one direction, we must reduce our demands in another. Atmospheric pollution, for example, can be intensified only so much before it endangers life and thus foreshortens the limit to population size. This space of operations, rigidly enclosed by the universal limits, but itself being flexible, contains all the options for the future, both conservationist and nonconservationist.

5. INSTITUTIONS AND THROUGHPUT

Institutions define and delineate collective social action and political response. Our lives at the personal, local, national, and international levels are molded by our institutions; collectively we mold these.

The family, the neighborhood, the school, or the church at one level, and the labor union, the profit-making firm, the cooperative, at another level, are all "institutions." At yet another level, the automobile, the dishwasher, color TV, disposable diapers are "institutions" in North America today, as too are advertising, a two- or three- or four-week vacation, the "restricted" right to carry firearms, and any and all other aspects of our lives which we have come to accept, rely upon, consider "ours as of right."

In very broad terms, the two principal institutions in any modern society which are the prime regulators of production, distribution, consumption, and, by implication, of throughput are the market and/or the state.

Does the market system—as it is currently structured in North America—operate efficiently and effectively from society's viewpoint? Does it allocate resources in a meaningful way? Does it do this over space and time? Can it take into account the needs and expectations of generations yet unborn? If the market does not perform such functions well can the state?

To relate these and a variety of cognate questions to our basic model of industrial throughput, it is useful to regard institutions as information processors, that is, instrumentalities which mediate between subjective needs and objective resources. Institutions, however, process information and react to stimuli in a preprogrammed fashion. It is this preprogramming which may be either highly efficient or highly wasteful.

Institutions are intermediaries. Needs feed into the institutional nexus and are filtered by the appropriate societal screening techniques. For example, in a free-market economy only desires backed by purchasing power become demands. Here, if it is allowed to operate without interference, the price mechanism acts as the damper. Through the price mechanism in a free market, or through regulation in a planned economy, certain resources *only* are selected for industrial processing. With appropriate dampers on both sides, supply will equal demand and the institutional system will be in equilibrium. There will be neither surpluses nor deficits.

Frequently institutions themselves can disrupt this balance. Institutions, in other words, may feed back on needs and amplify them. This is what can occur, for example, as a result of certain types of advertising in a free-market economy. Consumers are encouraged, indeed often cajoled, into buying more and more of this or that and at any cost. Such amplification can lead to increased resource use, which in turn can result in runaway industrial growth, as resources are used up in ever-increasing quantities and at ever-increasing rates to meet amplified demands.

Economic and social systems, fueled by such "positive feedback loops," may find the stresses these impose unbearable. At the limit, these may eventually collapse. This is the Malthusian solution. To avert collapse, institutions must adapt and adjust to changed circumstances and, in the process, change the circumstances under which they themselves operate. It is most important therefore that the information our institutions process be as complete as possible, that the social costs of growth, as well as its benefits, be considered.

Many institutional strains and stresses affect throughput. We link them to the growth of cities, and to their effects on surrounding areas.

There are signs everywhere today of an explosion in the growth of cities and their surrounding metropolitan areas. Between 1920 and 1977 urban populations throughout the world more than tripled, while rural populations grew by only one-third. Historically, cities have housed, clothed, fed, and in other ways provided for only a very small fraction of any total population. This reversal, largely brought about during the last hundred years, is, we may speculate, a direct consequence of the application of science and technology to the solution of the pressing human needs for food, clothing, housing, employment, excitement, stimulation, and other manifold objective and subjective desires.

Urbanization on a global scale is not only very recent but likely to accelerate in the future. Given current world demographic patterns, which are already set in place, it will be almost impossible to reverse the trend. By the end of this century it is more than likely that there will be vast concentrations of urban populations connected via networks throughout much of North America, Europe, Latin America, and the Near and Far East. Huge conurbations are anticipated between New York and Washington and between San Francisco and Los Angeles in the United States. Mexico City, for example, will have a population exceeding 21 million people by A.D. 2001, simply as a result of natural reproduction rates and movements already set in place. Likewise, Canada will not escape these centralizing tendencies.

If there are no fundamental changes in our collective attitudes toward the benefits to be derived from living in cities, the only kind of daily life which most Americans and Canadians in A.D. 2001 will know will be urban life. For example, out of a possible future Canadian population of 32 million in 2001, sixty percent will live in *only* twenty-two metropolitan areas. Moreover, Toronto, Montreal, and Vancouver will accommodate nearly half of this total, or, in round figures, perhaps 10 million. Similar, if less dramatic, predictions can be made for the United States.

Such expectations must give pause to even the most optimistic real-estate broker or land speculator, let alone to those future municipal and/or provincial administrators who will have the job of providing the range of public goods and services that cities of such size will undoubtedly require.

Urbanization on a scale such as is envisioned for that time is *now* in dynamic process. Its effects are far from clear and must be the subject of much speculation. We can, however, identify certain characteristics and problems.

Three main stages of urban growth are usually identified:

1. The initial concentration of certain social, cultural, political, and, consequently, economic activities in specific geographic localities as a consequence of past military, government, natural-resources-based, or commercial activities.
2. The implantation of significant public and/or private infrastructure to further support and/or regulate existing activities, and to expand upon these, or to develop new ones.
3. The emergence of megalopoles, areas with high population density and concentration of economic and social activities. This concentration first occurs in the urban core, next in surrounding industrial belts, and ultimately in "sleeper suburbs"—areas frequently several miles distant from where most inhabitants earn their livelihoods.

As a city develops, creative talent flourishes, adaptation to change takes place, and specialization allows for an enlargement of markets. The quality of public and private information, especially that which is technologically based, is enriched and increased. This information becomes widely diffused and stimulates further change. Producers find they can reduce their costs because they can rely upon more, and higher-quality, information. The latter, moreover, becomes a key to consumers' knowledge about, and choice among, an increasing range of alternative goods and services.

In the second and third stages of urban growth, but particularly in the second, a variety of economies become available to individual producers, economies which are external to, but enjoyed by the individual producing unit. These range from the creation of a social atmosphere supportive of innovation and technical change through the consequent creation of a skilled and educated work force, to the provision of a variety of public (collective) goods and services—hospitals, transportation networks, water supplies, sewage facilities—which can be enjoyed by the individual producing unit but for which the costs are assumed by society at large. Indeed, the growth processes of urban areas could largely be identified in terms of such economies.

Although external economies favor, and are favored by, larger cities, by the same token big cities may suffer from external diseconomies, not necessarily present in their smaller counterparts. External diseconomies begin to appear as soon as a city becomes "too big," are evident to us all, and manifest themselves in a variety of ways.

Direct economic costs arise as a result of the provision, at the municipal level, of a variety of public services which might not otherwise be required, e.g., large police forces, public housing and recreational facilities, child-care centers, welfare services, and clinics of various types.

Environmental costs arise as a result of air, noise, and water pollution, effluent discharges of all types, high-energy consumption rates per capita, aesthetic costs, and so on.

Personal costs result from higher incidences of certain diseases in cities, especially those which are stress-induced or of the respiratory type, from traffic accidents, higher crime rates, perhaps as a result of general alienation and loneliness.

In practice, it is difficult to distinguish among the above in any precise way. The effect tends to be cumulative. Thus, the provision of housing and welfare services, schools, public utilities, and other public services in the urban core presents for major cities around the world an increasing social and economic dilemma. New York City is almost bankrupt. Its tax base has been eroded as a result of the flight of people, capital, and industry to the suburbs, or to other parts. London and several other major cities are experiencing, or likely will experience, similar situations. Montreal, Boston, and Detroit are unlikely to be spared these problems in the future, if present trends continue.

In their attempts to escape from high taxes (imposed to support a variety of public services) as well as inadequate or very high-priced housing, congestion, stress, air and water pollution, crime, and the fifty-seven other varieties of inner-city disease, people and industry flee the central core, thus exacerbating the problems not only of the city but of the suburbs too.

There are a variety of costs to the city associated with this process. These range from the need for more highway construction and the consequent "shortage" of available land which this might entail, to the heavier use of certain roads at certain peak periods, with its concomitant frustration and personal health costs; to fruitless searches for parking spaces; to time wasted in commuting from A to B; to those many costs which arise out of the need to light, heat, and otherwise maintain the urban core, even though it might largely be depopulated at night when its work force returns to the suburbs. Other costs associated with both urban concentration or sprawl defy neat categorization, but are no less real.

Although it is now generally well recognized that the levels of postindustrial pollution and postconsumer waste tend to increase with, and faster than, the size of cities, it has not until recently been as well recognized that modern cities are no longer confined by urban space. Indeed, the growth of cities has led to their gobbling up the immediate, and often distant, environments and to the imposition upon these areas of a particular, and often inappropriate, character.

Climatic and geological conditions in Canada make the same land desirable for both agriculture and human habitation. Most Canadians, therefore, prefer and (given the present structure of cities in a narrow band close to the U.S. border) are almost forced to live upon some of the best agricultural land.

Between 1961 and 1971, 370,000 acres of prime agricultural land were surrendered to the growth of suburban housing estates in Canada. This land was capable of feeding 23,000 people. This loss may not seem like much in comparison with the shelter provided, and the communities built up, and it probably is not. However, with increasing energy costs, limits upon the productivity of marginal land are being increasingly imposed. It is likely that the severity of these constraints will become more evident as time passes. Thus, expansion of Montreal onto its fertile agricultural plain; the implantation of an international airport outside of Toronto (Pickering), which so far has destroyed 40,000 acres of excellent land, although the airport has not yet been opened; or the diminution of the forest zone in the Niagara peninsula, all should give cause for concern. These represent costs which result from urban sprawl, and any proper accounting must consider these costs.

Cities concentrate, rather than diversify, service activities. Moreover, they replicate these at an ever-increasing rate. The implantation of banking, financial, insurance, or government institutions in major cities leads, via a circular process, to further growth of these institutions. The revenues of banks or insurance companies, for example, are today frequently invested in urban financial centers, often in commercial real estate, office towers, and so forth. In order to protect their investments, such large financial institutions frequently must, and do, resist actions which might decongest urban centers, for to support such action would be, ultimately, to reduce the value of their holdings (that is, their land values).

Nor is the municipal mayor, faced with the prospect of raising the operating budget largely from real-estate-based taxes, likely to encourage movements toward decentralization. The opposite more frequently is true. This is especially so if a higher assessment is made on business enterprises than upon homes.

There is a final category of costs which should be considered, this time in the spatial context: namely, those which arise as a result of regional "underdevelopment." The neglect of growth in certain areas presents its own particular set, or sets, of economic and social consequences.

If, for example, there is a small population in a certain area, or if the population is declining, the public services already in situ may cost more per capita to maintain than will the equivalent set in a larger urban area. This is so because, with the provision of any type of public good or service, there normally are certain minimum fixed costs. These will decline in proportion to the frequency with which the service is used. Thus, if there are fewer people to use the services of a community nurse, say, who herself has to travel longer distances between patients, the per-capita cost of maintaining that nurse may be higher than the equivalent per capita cost of maintaining her big-city counterpart.

It may be true, of course, that the rural and urban nurses do not provide exactly the same kinds of services and that, therefore, it becomes very difficult to measure the relevant per capita use costs. Equally, in smaller areas, the *need* for police, welfare, and other social services may be less than in the city.

It may also be true that the relative poverty, at least in monetary terms, of most underdeveloped regions in the United States and Canada will mean significant limits *are* imposed on such regions' abilities to raise public revenue and, thereby, to support those kinds of public investments which may be necessary to move their inhabitants beyond their current poverty. For those who choose to stay in the rural areas this may mean present lack of educational opportunities and a future lack of employment prospects; a situation which, cumulatively, may lead to further decline, perhaps of critical proportions, in the number of those choosing to remain in the rural areas. The redistribution of national wealth—through the variety of transfer-payment mechanisms which now exist—can of course alleviate this situation. Again, however, there are many costs in this process, most of which are rarely spelled out in detail.

In the light of the above, whereas it is true that the fisherman in Louisiana may have a lower per-capita income than the shop assistant in Los Angeles, does this mean that he is necessarily any worse off in real terms? Why is income generated in Los Angeles? Because of the city's internal economic growth. But what might much of this growth, and its concomitant employment, be ascribed to?

Automobiles certainly increase Los Angeles's revenue; these generate repairs, use gasoline, require the construction of highways, parking lots, and so forth. Similarly, psychiatrists' incomes may be high in Hollywood but is this an indicator of Los Angeles's well-being, or lack of it? Similarly, finding ways of coping with postconsumer pollution and waste undoubtedly gives garbage collectors in New York higher incomes than some schoolteachers in Kentucky. But again, is this an indicator of wealth, or welfare, and are these necessarily the same?

In all of these examples the essential conflict is one which opposes liberty to order and juxtaposes the interests of the individual to those of the collectivity. It is appropriate, therefore, that we return briefly to the market and the state as regulators of industrial throughput.

A complete analysis of the objectives of any one firm is a

complex undertaking. That for several firms, for industries, and for economies is much more so. In any discussion of the market mechanism and the way it regulates industrial throughput, three distinctions must be kept in mind: the distinction between center and peripheral firms; the distinction between competitive and monopolistic market environments; and the distinction between consumer-goods and industrial-goods markets and producers.

Center firms are the large dominant producers and/or consumers of industrial goods and services (large national and multinational corporations, mining companies, AT&T, Exxon, General Electric, General Motors). They *may* sell to, or buy directly from, the public, but often they sell or buy through wholesalers, distributors, and retailers. In this regard they can face sophisticated buyers and, perhaps, less-sophisticated sellers. They have large capital resources, often finance activities out of retained earnings, and frequently produce a host of complementary and competing goods and services simultaneously. Thus, it is difficult to identify with precision exactly which part of any of these operations is costing exactly what and to whom. Furthermore, tacit agreements, at least over such items as market share or joint or subsidiary ventures are frequently encountered.

Peripheral firms are the smaller, less-structured remainder of producers and/or consumers. Market conditions here range from monopoly to effective if not perfect competition. Such firms generally are price (and therefore cost) takers rather than price and/or cost makers, as are more frequently the center firms.

Similar structures exist in labor markets, between highly organized "union-protected employees" and their unorganized "non-protected" counterparts.

It is against this complex background that any assessment of the market mechanism must be made.

Under free market conditions, a nonrenewable natural resource derives its market value from the prospect of its extraction and sale. If the free market is also competitive, owners of nonrenewable resources could expect two things to happen: the prices they could sell their resources for would increase as the supply diminished or the demand increased; the value of their capital assets, i.e., the nonrenewable resource still left in the ground, would also appreciate, and at a rate equal to that of any equal-risk investment.

These two conditions define a truly exhaustible resource. Implied here is the notion that both the stock (asset) and flow aspects of nonrenewable resources are interwoven. According to R. M. Solow, if these two aspects have been perfectly harmonized through the "operations of futures markets . . . the last ton [of any resource] produced will also be the last ton in the ground." In other words, "the resource will be exhausted at the instant it has priced itself out of the market."*

These equilibrium conditions in the pricing of nonrenewable resources by the market mechanism leave many questions unanswered. In particular, the stock and flow aspects of nonrenewable resource markets may give contradictory signals to resource owners or users (in our terms, provide misinformation). Unless net resource-transaction prices are growing at a rate at least equivalent to the prevailing compound-interest rate from similar risk-class ventures, resource holders will be tempted to extract and sell as much of their available resources as possible, rather than to hold on to them. The consequence of this will be that as more resources are placed on the market the received prices for these will be lower, thus stimulating further the use of non-renewables. A myopic, short-term view may, therefore, exacerbate resource scarcities over the longer term.

The opposite may also be true; if resource owners anticipate significant asset appreciation they will be tempted not to draw

*R. M. Solow, "The Economics of Resources and the Resources of Economics," *American Economic Review* (May 1974).

down from reserves to meet current demands, the effect of which will be to drive up resource prices further. This situation can, again, lead to disequilibrium: the expectation of even higher prices, and the supply restrictions thereby stimulated, can lead to these higher prices becoming a reality.

There are these further complicating factors in the pricing of a nonrenewable resource:

First, futures markets for resources are not well established, and, in any case, are fraught with imperfections, not least of which is inadequate information. Neither companies nor governments like to let their potential competitors know exactly what their reserves of nonrenewable resources are. Frequently they don't know themselves. Without such information, however, futures markets cannot function effectively. It is difficult to discount stocks at appropriate rates if these stocks are not known in the first instance.

Second, even if futures markets were well established, they would still be subject to a variety of surprises and shocks. There are surprises resulting from new technological applications for certain resources, and, not least, from political events. The recent success of the OPEC cartel in jacking up, and subsequently maintaining, world oil prices at an "artificially" high level is, of course, the most obvious illustration.

Finally, the concentration of resource ownership in the hands of private, public, or parapublic monopolies significantly influences the working of resource markets. In this context, the love-hate relationships that frequently exist between multinational corporations (which may derive their revenues largely from resource extraction and processing—such as ARAMCO in Saudi Arabia) and their host nations can further distort international market pricing.

Multinational corporations can bring, and certainly have brought, capital to resource-rich but capital-poor nations. Additionally, such corporations have greatly facilitated the international transfer of technology. However, many host nations have found such benefits to be outweighed by the costs; these latter being represented as the exhaustion of the host nation's natural resources, the internal political power exerted by the corporations over domestic affairs, and the creation of dual economies, i.e., enclaves of affluence surrounded by seas of poverty.

With respect to the pricing of renewable, including human, resources, major criticisms which can be leveled against the market mechanism are, as with nonrenewables, difficulties in handling longer-term questions and a neglect of "externalities." In addition, distortions are introduced into pricing as a direct consequence of either monopolistic practices or the activities of regulatory agencies.

We have already referred to external diseconomies in our discussion of the social costs of urbanization. With respect to what might be called "long-term structural change," the question of whether the market mechanism is an appropriate vehicle is confusing and complex. To illustrate: The construction of the transCanadian railroad was a massive undertaking. It came about principally at the initiative of British Columbia as a price for entering Canadian confederation. If a cost-benefit analysis had been undertaken in the early 1870s, say, and if the only item entered into the calculus then had been the prospect for immediate profit, it is questionable whether the railroad would have been built as early as it was or in the same manner.

In other words, industrial practices change (in response to the changes in the environment which embrace these) not instantly but rather only after a time lag.

Time lags may be substantial, depending on how long it takes to create new technology and, perhaps more importantly, on the time needed to recoup the investment made in response to the old demands of the society. Moreover, the increasing complexity and sophistication of industrial practices today reinforce any temporal rigidities which there might be in industrial structures. Companies these days are increasingly dependent on one another, as all are upon government. Thus any "switch-over time", which might be "optimal" from a social viewpoint might be shorter than that dictated by existing market conditions and constraints. In the transCanadian railroad example, therefore, the investment made at that time in wagons and water-borne transportation was a constraint on railway construction. The private transportation firm had an investment in an old form of technology. Government thus had to assist in the movement to the newer, that is, railroad, technology.

As far as monopolistic practices are concerned, a variety have always existed, and likely will continue to exist. Frequently governments have to intervene in markets to police their operation. However, since the end of the Second World War a variety of events have occurred which have tended to blur the traditional concepts and constraints of markets and the marketplace. Today, for example, IBM and Xerox battle for office copy and duplication hegemony while IBM also struggles with AT&T for control over the electronic transmission of data. One also finds new industries eventually being able to support only a few large competitors, with other corporate entrants, themselves giant entities in different lines of business, forced to withdraw. Appliances and computers are but two of the areas where this struggle of "colliding collectivities" has occurred.*

The state is an alternative to the market as a mechanism through which to regulate industrial throughput. As *the* central planning body, the state may act without guidance from the price signal and initiate activities which may not now be profitable (environmental-protection industries) but which may serve the public interest.

---

*We thank Stanley Shapiro for this term and for several of the ideas developed here. Interested readers are referred to his recent paper, "Marketing in a Conserver Society," *Business Horizons,* April 1978.

The state can also discontinue or forbid profitable activities if these are found to be contrary to the public interest. Precisely because it is *not uniquely* constrained by the profit motive, the state, it is sometimes held, may be able to take a longer-term view in its planning and in the execution of its activities than is possible for the individual or the corporation acting through the market system.

Removed from market constraints, and the discipline this can sometimes provide, governments can become, and frequently seem to have become, inefficient and wasteful not only of taxpayers' monies, but also of taxpayers' time, energy, and efforts. Indeed, much of the immense growth in government activity of recent years has occurred without precise criteria or raison d'être being first established.

Today, we find a variety of constituencies struggling for control over both the administration and the enjoyment of the benefits associated with respective government departments or agencies. Who, in many such organizations, should take the ultimate responsibility for past, present, and future actions frequently remains unclear. Indeed, it is often undefined. Consequently, the buck does not stop but rather gets passed to higher or lower or farther or nearer echelons. Information not only gets processed but reprocessed and then processed again. The net result, of course, is that the information is scrambled, signals are distorted, and politicians frequently make decisions *not* in response to adequate information but in response to situations of information overload. We are all familiar with such bureaucratic pollution and its consequences.

Government is not unique in this. In large corporations too the buck is passed from A to B to C. Decisions do get made, but rarely are they reached by the dynamic entrepreneur acting on a hunch, as may have been true in an earlier era. Now decisions are made by management committees, with consensus as a major objective. Where, within the institutional context, might we look for reasons for, and answers to, many of the current problems we face?

In the chapters that follow we suggest a variety of techniques and approaches to overcome the present institutional constraints to sane resource use and effective postproduction and postconsumer waste management practices. This is an urgent matter. The disposability which our past affluence may have permitted us, and the resource profligacy this has encouraged, will not stand us in good stead in a resource-scarce and energy-expensive future.

II
CONSERVER
SOCIETY ONE

Growth with Conservation

World View:
Efficiency and Expansion

Motto:
Do More with Less

6. WHAT IS A CONSERVER SOCIETY?

The basic idea behind the conserver society is the notion of conservation. Webster defines "conservation" as "preserving, guarding, or protecting: a keeping in a safe or entire state." A decision to conserve implies that future consumption is valued more highly than present consumption. Given the uncertainty of things to come, it may be assumed that most people prefer present to future consumption. This is true in most instances but there are exceptions: a resource which is abundant now but is expected to be scarce in the future may be suitable for conservation. The time preference which emerges from a psychological comparison of present as against future consumption is at the basis of a decision to conserve or not to conserve.

In order to make this notion of time preference more specific, we can define conservation as the process of prolonging (either by preserving or by using-and-recycling) the useful life of resources.

A resource is anything we subjectively feel has the capacity to satisfy our needs or wants. It may be concrete or abstract, material or spiritual, raw or transformed. Useful life refers to the time during which a resource is indeed capable of satisfying. Resources do not die, but consumed ice cream changes its state and cannot continue to produce the initial satisfaction associated with eating it.

The distinction between conservation by preservation and

conservation by use and recycling is important; some conservationist scenarios are based on the preservation theme whereas others stress recycling. (In both situations the stated object of prolonging the useful life of resources is realized.)

If conservation is thought of merely as non-use, it is tempting to suppose that the only perfect conserver would be a corpse. But a live consumer may also be very much a conserver. If he or she consumes only fully renewable resources and there are no entropy or other irrecoverable losses, he or she is a perfect conserver.

Nor is it necessary that, for a conserver society to exist, all its members be conservers. It is theoretically possible for some segment of the population to conserve by preserving, another to recycle, and yet another not to conserve at all. To argue that a conserver society can exist only if every single member becomes a conserver is to commit the fallacy of division, which mistakenly assumes that what is true for the whole is necessarily true for the parts. Notwithstanding this technical point, in an ideal conserver society, most members should conserve.

The conserver society idea, however, is richer than the mere concept of conservation. In a more subtle sense we define a conserver society as a societal organization in which high priority is placed on the objectives of: waste reduction in the throughput process, greater harmony with nature, and a longer-time horizon as a basis for decision making.

"Waste" is a highly relative concept which depends on the intended outcomes of a process as well as on the perceived scarcity or abundance of the resources to be used. In the nineteenth century, when industrialization was beginning, stretches of virgin land were referred to as "wastelands." The remainder of that land, now viewed as green space and threatened by industrialization, is deemed in danger of being "wasted." Obviously an understanding of waste rests on value judgments and is subjective. Even assuming a consensus, the eradication of waste in the throughput process is probably not possible. What can be achieved is waste reduction—by substituting for any existing process another which is more efficient in conserving scarce resources.

The second condition, greater harmony with nature, is justified not only from the point of view of the preservation of the biosphere but also from the selfish standpoint of the preservation of the human species. Even if we see ourselves as parasites upon nature (a notion that some of us entertain), what we most want to avoid is the death of our host. For our own good we must respect the ecosystem's balance.

This respect should not, however, be taken to ridiculous extremes. A man-made project which slightly alters the environmental picture is not necessarily bad per se. We must avoid the fallacy of ultra-stability. The biosphere, to survive, must adjust to new conditions, including those brought about by the human species. What "harmony with nature" and "environmental respect" really mean is the careful assessment of the probable and possible effects of our decisions *before* we make them. This ties in with the third minimum condition, which advocates a longer time horizon in decision making.

When economists attempt to measure the possibilities of success of a project, they use a cost-benefit calculus: all real costs whether visible or hidden are accounted for and compared with all the expected benefits. If the overall balance is positive, the project is worthwhile.

However, let us note that it really all depends on what time horizon is used. Many projects that will show only losses in a three- to five-year time span will fully realize their potential in ten to fifteen years. Therefore, a near-sighted cost-benefit calculus may yield misleading results either on the cost or on the benefit side.

The game of escalation of time horizon is often played to an amusing extent among futurists. In one "futures" bookshop four books were displayed in a row. The first, a conservative business-cycle report, was entitled *The Next Two Years: the Economy Recovers*. The second was the proceedings of the World Future Society's convention and contained an anthology of various articles on futuristic issues. It was entitled *The Next Twenty-Five Years*. The third was the Hudson Institute's more ambitious *The Next Two Hundred Years*. The last was the book that dwarfed the other three: its title read unabashedly *The Next Ten Thousand Years*.

Strangely, this last book could prove to be more accurate. The reason is simple: whereas the first three dealt with volatile social issues, with the unpredictable human element at their center, the last dealt with astronomical issues. It is paradoxical but true that it is far easier to predict a total eclipse of the sun in the next 350 years than your own mood tomorrow morning.

Predictability and unpredictability set aside, the human being is not equipped to think in the very long term. Witness, for instance, the doomsday prophet who gave a lecture about the gradual weakening of the sun's rays. "In 500 million years," he expounded gravely, "the sun will be dead." At which an anxious young man broke into a cold sweat and trembling went forward to ask: "Excuse me, sir, how long did you say the sun would last?" The doomsday prophet responded peremptorily: "500 million years." "Oh," said the young man, suddenly greatly relieved, "for a moment I thought you said 50 million years."

Above and beyond the statement of broad objectives, it is necessary to spell out in greater detail the assumptions behind the notion of a conserver society.

Assumption 1: Conservation is a means to an end, not an end in itself. The end is human fulfillment in harmony with nature.

Very few utopias in literature see conservation as an end in itself. To decree arbitrarily that it is, is tantamount to giving infinite value to the future and none to the present. This is hardly acceptable. It is more reasonable to assume that conservation is one means among many to ensure a high quality of life for this and future generations.

Assumption 2: Human happiness depends on a balance between needs and commodities.

A need may be defined as a situation of want, whether subjectively or objectively induced, which, if not satisfied, is a source of discomfort to the person experiencing it. A commodity is a need satisfier. Needs are what psychology is all about. Commodities are what economics is all about. In the words of W. L. Gardiner, Happiness is a commodity for every true need.* This is not an empty statement. It means that the conserver society must take into account the needs/commodity interface and deal with it adequately.

Assumption 3: In a throughput process it is not possible to "economize" all inputs simultaneously.

This is a variant of a proposition in production theory stating that whenever we use less of one input (vis-à-vis an alternate production mode) we use more of at least one other. All inputs cannot be economized (used sparingly) simultaneously if production is increased. This is especially true when a full list of inputs is made. Washing clothes by hand saves energy but uses up more time than machine washing. Overworking may save electrical energy but interfere with health. Saving time, energy, and health may be possible by employing *more* know-how or

---

*Coined by W. L. Gardiner, in the psychology paper of the GAMMA Report.

information. It is therefore necessary to select carefully the inputs we are going to economize.

If, in fact, we cannot economize everything simultaneously, we must devise selection rules to choose what it is we wish to conserve.

Assumption 4: There are many selection criteria.

Before the selection of inputs to be preserved (or conserved via use-and-recycling) the following questions must be at least posed: (a) *What* do we wish to conserve? (b) *For how long* a period? (c) Who are the *beneficiaries* of this conservation and who are the *losers,* if any?

To provide answers, selection criteria based either on economic, environmental, or value considerations, or a combination of all three, may be put forth.

Conservation for purely economic reasons involves the selection of inputs which are either already scarce or liable to become so in the near future. The rate of conservation and its costs/ benefits will depend on rates of discount applied now and in the future. It is understood that in a perfect economic system scarcity would be reflected in the price structure and all items would be fully costed.

Conservation for purely environmental reasons involves the selection of inputs which must be conserved to retain the viability of the ecosystem, which depends, at least partially, on "requisite variety."

Conservation for value-dominated reasons requires a full explication of the value system. For instance, we may want to conserve sacred cows, a species of wildflower, a landscape, a Norman church, a language, a religion, or whatever.

It is obvious that only some combination of all of these criteria will be comprehensive enough to satisfy a discriminating public. This combination must somehow bring together the three sets of criteria, giving them a common denominator.

Assumption 5: There are several types of Conserver Societies.

To reflect the diversity of value systems and/or the urgency of the need to conserve, it is better to offer a range of options under the general heading "Conserver Society" rather than just one. We present three versions of the conserver society, which differ in degree of conservationist effect. Conserver Society Two (CS_2) is more conservationist than CS_1 and less conservationist than CS_3. Insofar as we claim that the status quo is CS_0, the other difference will be in terms of value change. CS_1 will be least distant from our present value system, CS_2 more so, and CS_3 most distant.

Assumption 6: The Conserver Society options must be both site specific and culture specific.

What this means is that the conserver society options must be ideally suited to their environment. In other words, Pasadena, California, Sioux Falls, Iowa, Contre-Coeur, Québec, and Come-by-Chance, Newfoundland, although all broadly speaking are on the same continent, do not face the same problems. Conservationist policies have to be regionalized.

Assumption 7: The Conserver Society must be made compatible with the goal of reducing income inequalities both nationally and internationally.

Environmental movements have often been accused of ignoring poverty and income inequalities and of indulging in overt Marie-Antoinettism. When the unfortunate queen of France recommended to the starving peasants that they should eat cake, she manifested the same insensitivities as those exhibited by extreme environmentalists who do not face up to the problem of poverty. Environmental Marie-Antoinettism advises workers who lack the necessities of life to go enjoy nature instead. "Let them eat sunsets," it is recommended. This type of argument is offensive to many low-income groups and Third World countries.

For this reason an ideal conserver society would take into account the problems of inequality and include it in its policy recommendations. As was pointed out before, the conserver society is not just "conservation," it is a comprehensive model for society which reflects the various interdependencies among social problems.

Sammy Squander Hears from his Old Friend Angus McThrift

Like many Canadians, Angus McThrift was of Scottish descent. He came from the same part of the country as the world's most famous (and tallest) living economist but had, early, given up the idealistic hope (and the nagging fear) of living in a society where his illustrious countryman's theories would be put into practice. Even as a teenager Angus had realized that being openly "sensible" with money only made a person unpopular. And Angus should have been popular. With his easy, friendly smile, and few, well-considered words, he seemed to have an inner strength. It showed in the light, sure movements of his compact athlete's body and in the reserves he seemed to have in times of crisis. There was nothing superfluous about Angus, but his friends couldn't love him unless he exhibited their own weaknesses: their recklessness, their cupidity, their easy-come, easy-go attitudes. There was no use saying, "I can't afford a cab, I'm saving up for a bicycle"; they would just sneer and leave him to go off alone on weekend cycling trips when, later, he finally acquired the basic machine. His friends' fancy bicycles were taken so much for granted that they lay in closets rusting and out of repair for years.

Because of this sort of thing, Angus spent a lot of time alone. He had borrowed a psychology textbook from the library and discovered that he was being subjected to something called "peer pressure"—apparently designed to make him conform. Obviously, unless he wanted to live a miserable life, a new strategy was needed.

He kept quiet. Through trying to save money Angus had discovered—almost accidentally—that it was very often possible to have a better time while spending less. He realized that people like Sammy Squander paid heavily for the privilege of not thinking, especially not thinking ahead. At least half of what they spent went for the dubious privilege of doing things or having them at the same time as everybody else. Just like sheep, they always bought the latest thing when it was obviously better to wait till the bugs were out of a new design and the spent market brought the price down. They would rush to buy new gadgets without shopping around, and they would never think of renting the item first, to see how well it did the job and how much it would really be used. When Angus tried to swap things with a friend instead of both buying new, so that they could have money on hand for a cheap off-season trip to the Bahamas, the idea was scorned. It seemed like too much bother.

To Angus, it seemed like too much bother to spend his precious leisure time and energies in department stores jostling frenetically in competition with hordes of other people, all trying to buy the same things. Nor did it seem worthwhile to suffer traffic tie-ups to get to overcrowded campsites and beaches where you couldn't see the sand for people. Early in his successful university career Angus planned ahead to do consulting work so that, once qualified, he could choose his own hours and take advantage of off-peak, off-season prices and pleasures. Thinking that a great many people could do the same, he lobbied energetically for "flexitime" in local government, industry, and business but was either ignored or put down as a crackpot student.

However, it was toward the end of Angus's college years that things began to change. He seemed to be becoming popular. The people he'd gone to high school with started coming to him for advice, and to borrow. After all, he was never short of money, never hassled for time; everything he possessed was of the finest quality and in excellent repair. He was in control of his life and could afford to be generous. As his friends became disenchanted with their closetfuls of cheap, obsolete goods (and with The System in general), Angus was finding out that being careful took less and less effort. It was simply a habit, as easy to adopt and as difficult to shake as any other. With this built-in advantage, Angus could relax and enjoy life.

This was the vision of the Canadian Scot being conjured up, not without envy, by Sammy Squander as he settled down to read Angus's letter from where he was vacationing on the French Riviera.

Le Lavandou, France
July 10th

Dear Sammy,

Blue azure sea, cloudless sky, the sun just setting on the Mediterranean, and, as I sip my pastis with a group of new-found friends, I am thinking of you back in Montreal and am dropping you a line. The atmosphere here is fantastic. Luckily the tourist hordes have not yet arrived (that is a doubtful pleasure I will miss since I'll be in Italy by the weekend of Bastille Day, when they start coming in). You won't believe it, but I've managed with about half the money I thought I would have to spend. First, the trip with Norlandic Airlines to Luxembourg was as lavish as first class yet, because it was a charter, cost exactly a third of what a regular flight would have cost. By being picked up at the airport by Francine (you remember Francine; she lives in Strasbourg now and came to meet me) I got a ride all the way to Paris. There I picked up my new car—and Sammy you won't believe it. It is a Renaud-Giarinni R-17. For openers, three roofs, one for the blizzard, one for the Riviera (a convertible top), and an in-between roof for chilly evenings. A 2 + 2, it seats four in great comfort. But this you'll find hard to believe. It accelerates faster than your Stratobird and at the 85 mph limit of European super highways it gets no less than 35 miles to the gallon. How's that? What's the secret? Efficiency, my dear fellow: a 1,500-cc engine with electronic fuel-injection combined with a five-speed gear box. That, with front-wheel drive, gives terrific performance yet excellent mileage.

Picture me then, dear Sammy, driving along on the beautiful mountain roads of the Riviera, the wind blowing my hair all over, stopping at little restaurants and above all taking my time. Lunch today was a menu-à-prix-fixe: soupe de poisson, terrine-

du-chef, paella valenciana, and mousse-au-chocolat. The price?
$3.50 with the tip. The secret? Look for little out-of-the-way
places far from the town centers. At night, Francine and I stay in
little two-star hotels, perfectly clean but no private bathroom—
who needs it anyway? It costs us about $6 for a nice room and
two breakfasts of café-au-lait and croissants.

My dear Sammy, I must stop now. Next time, let me manage
your finances and I'll take you along. Remember—waste not,
want not. A little careful planning and—la vie est belle.

Angus

P.S. I definitely broke up with Rita Righteous. Her austerity and
seriousness have turned me off completely. If she writes to you,
send her my . . . friendship.

7. THE SCOTCH GAMBIT

or The Value System of CS_1

Conserver Society One is designed specifically to change people's behavior without necessarily changing their value systems. The idea is to reinforce those elements in the existing value system that stress conservation, thrift, and economy coupled with progress, growth, and expansion. Behavior patterns are ways of doing things. There is more than one way of doing things compatible with a given value system and it is possible to reform value systems without altering the beliefs, ideals, and preferences that underlie them. That, in fact, is the assumption behind the strategy which in the next chapter we call RICH (Reform of Inefficient Consumption Habits).

But what parts of our present value systems stress conservation and thrift? The answer requires developing a stereotype which, like all stereotypes, is larger than life. It overemphasizes certain features, permitting the fuller exposition of an idea. Borrowing from chess parlance, let us call the value system of CS_1 the Scotch Gambit and describe its motto as "do more with less."

The traditional image of the Scot is that of a person who is highly effective in both thriving (prospering, accumulating, growing) and thrift (economy of resources, parsimony, husbanding). This cultural stereotype, although exaggerated, is nonetheless enlightening. It finds some confirmation in the more general analysis by Max Weber in *The Protestant Ethic and the Spirit of Capitalism.* Weber depicts an entrepreneur who is motivated by the Puritan ethic or its Calvinist variant. Whereas the medieval Catholic Church advocated poverty and resignation to life's ills, the Protestant Reformation glorified initiative, progress, development, and economic growth. The parable of the talents, a metaphor from the Gospels and a mainstay of Protestant thought, showed that, of the three sons who each received a talent from his father, only one did the "right thing" with it. The first spent it, the second buried it in the garden, but the third, the blessed one, invested the money and made it grow.

The resulting Protestant ethic, culturally strong in Scotland, became an ethic of high economic growth and high efficiency. The great North American entrepreneurs of Scottish background fall very much within that stereotype. These include Carnegie, Molson, Strathcona—all output maximizers ("growth maniacs," as it were), who yet, by being input minimizers, were also conservers. Whereas in the Catholic tradition it was deemed to be more difficult for a rich man to go to heaven than "for a camel to pass through the eye of a needle," in the Puritan ethic God showered his blessings on the rich man and rewarded him for his diligence and hard work. It was not infrequent for the nineteenth-century Puritan entrepreneurs to declare that, in their businesses, God was their "senior partner."

But the CS_1 entrepreneur is more than just a Weberian prototype. He is not only prudent, he is also dynamic. This is where we shift from Max Weber to Joseph Schumpeter. According to Schumpeter, the entrepreneur has three desirable qualities: management skills, willingness to take risks, and an innovative temperament.

Schumpeter argues that the main agent of technological change is the capitalist entrepreneur, who, ideally, acts as a spur to invention as well as innovation. He knows how to translate science into commercial successes. He is a careful manager of resources and knows how to calculate risks.

The Schumpeterian entrepreneur thrives on the business cycle, which becomes the chief instrument of prudent resource management. At every downturn of the business cycle, the inefficient marginal producer goes under; only the strong, healthy, innovative business survives. The entrepreneur saves his firm by becoming more efficient. All those firms that cannot adapt go out of business. The business cycle then trims the fat off the economy, and, in so doing, is a conservationist influence.

The Schumpeterian system is thus the epitome of the high-growth, high-conservation market ideal and, together with Adam Smith's Invisible Hand and the Weberian entrepreneur's prudence, sums up the case for a market-led conserver society.

As an alternative to the entrepreneurial market-led conserver society, a statist version is possible. This would derive its inspiration from European and particularly French schools of thought.

As is well known, the "laissez-faire" British liberal economy never completely convinced continental policy makers. The economic and political history of Europe since the Industrial Revolution is one long sequence of what the French call *dirigisme* or state-controlled planning—with occasional interludes of liberalism. In state planning the conditions for efficiency may be met, not by the competitive mechanisms of the market, but by the strong, guiding and very visible hand of the central government.

Efficiency-directed central planning may take many ideological forms. In the French tradition of classical mercantilism, production was maximized in order to ensure more power to the state. This was Colbert's mercantilism in the seventeenth century and it had echoes in de Gaulle's Fifth Republic.

Both the Scotch and statist dimensions of CS_1, although they differ in basic value assumptions, are centered around the same goal: efficiency.

Values involve not only man-society relations (as we have described in the Scotch and statist ideas) but also man-nature relations. The CS_1 model could be described as enlightened anthropocentrism in terms of the man-nature relationship. Ecologists distinguish between an autecological analysis of an ecosystem (the ecosystem viewed from the vantage point of one species) and a synecological analysis (the ecosystem viewed from the point of view of its own general interests). The CS_1 scenario adopts an autecological approach, stressing human interests but, unlike some present social practices, does not have a strong man-over-nature bias. Instead it has a man-with-nature bias, the emphasis being on symbiosis. Man is the beneficiary, but the biosphere is not the loser. The philosophy of CS_1 is nothing more nor less than the philosophy of efficiency, economy, and intelligent hard work. The caricatured Scotsman can don the garb of the free-market entrepreneur or that of the innovative public servant.

"Doing more with less" can be a cliché, an empty slogan. To give it flesh, give it meaning, give it significance, one must translate it into strategies of implementation. At the back of our minds must continually lurk the question "more of what with less of what?," which is another way of restating the fundamental problem of conservation. "Doing more with less" will result in "getting more with less": more of the "goodies" (better quality of life, higher standard of living, more *joie de vivre*) with less of the "baddies" (premature resource depletion, pollution, ugliness, disease).

Let us now look at the six principal strategies to achieve CS_1: RICH, Sharing by Renting, the Management of Time, Conserver Technologies, Full-Cost Pricing, and the Optimum-Mix Economy.

8. RICH

or The Reform of Inefficient Consumption Habits

It is not accidental that "Reform of Inefficient Consumption Habits" yields the acronym RICH. It implies that considerable benefits, material and nonmaterial, can accrue from reduction of consumer waste. This does not involve significantly altering our life-styles or even reducing anybody's current standard of living. We simply weed out those inefficient, unproductive, and unnecessary consumption practices that we have adopted during the era of cheap energy and apparently limitless raw materials. RICH attacks the kind of extravagance that costs us dearly without contributing much to our comfort or convenience.

Changes in Western industrialized society are generally brought about not by the independent decisions of its individual members but by those of key decision makers in some complex bureaucracy. Most problems, however, which beset our society encroach gradually and are caused by the day-to-day behavior of individuals. They can be solved, in the final analysis then, only by a change in that behavior.

The global troubles described in such publications as *Limits to Growth* and *Small Is Beautiful* have captured public attention but, so far, have not significantly changed consumption practices. Perhaps the reason is that the situation, when depicted on a world-wide scale, seems much too complex and remote to offer obvious indications for change in people's lives. The temptation is to think, "Somebody else, probably the government, is doing something about it." It should be remembered,

however, that governments are complex and unwieldy institutions, which move only at public demand and then only with
agonizing slowness. Government bureaucracies are also constrained by the interests of political parties, which rarely think
beyond the next election. Waiting out long bureaucratic delays
allows a problem to grow more acute, or a crisis to pass, if pass
it will.

One of the crises that will not pass is the ever-widening gap
between our demands on the physical environment and that
environment's ability to support those demands. The gap is
widening because as we intensify our use of nonrenewable resources the supply, inevitably, is running down; as we overex-
tend our renewable resources the systems by which they regen-
erate themselves are gradually being paralyzed; as we blithely
dump dangerous pollutants into the soil on which we grow our
food, into the water which we drink, and into the air which we
breathe, we may be desecrating them beyond the ability of
clean-up campaigns to save the environment.

We can go a long way toward reducing or avoiding these
problems simply by cutting down on waste, without altering
fundamentally the way in which we live. No one would suggest,
for example, that the bright lighting of empty rooms caters to
real human needs. Moreover—and the important point in this
context—it is much less wasteful for individuals to decide by
themselves to turn off the lights than to set in motion the complex and sluggish machinery which will persuade them to do so,
which will elect a political party on a platform of energy conservation, which will then employ civil servants to devise economic incentives, which will then encourage individuals to turn
off lights in empty rooms.

That part of throughput which is wasteful is kept going in
tandem with the portion which actually provides consumers
with satisfaction. Waste is empowered to move around the
throughput cycle not only by those who profit from it but also by

the consumers who are careless enough to allow it to happen. Unnecessary products are manufactured along with the necessary ones. The two are often purchased together. The glaring difference comes at the points of use and disposal. Products which are virtually useless are, logically enough, little used. Their main purpose is in being discarded; their only value, nuisance. The latter is not entirely true, as can be seen by considering the most glaring example of waste in our society—overpackaging. In order to buy some food products we are also obliged to purchase elaborate, brightly colored, expensive, and often unnecessary paper, plastic, and metal containers. Their value lies primarily in persuading us to buy the product. It is debatable whether this "value" does more for the producer or for the consumer who pays for it. As soon as money or credit card changes hands, the fancy package has completed its usefulness. If it was an asset at the check-out counter, it may soon become a liability to its purchaser and to society.

This is especially true if the package contains food or other items which are used immediately after purchase. The package quickly becomes dead weight, which, at its most innocuous, takes up space. At worst, the chemical dyes which endow its colorful appeal will pollute the soil on which it is dumped; the complex structure of the plastics or metal alloys which give it "rigidity" will be a virtually indestructible eyesore. It is removed from the consumer's immediate vicinity (and awareness) only by virtue of high taxes paid for garbage disposal. Why, it might be asked, do consumers prefer this kind of packaging?

As usual, the answer can be given only by a chicken-and-egg explanation. First, the package is attractive—it appeals to humanity's very primitive desire, somewhat akin to the magpie's, for bright prettiness. If you imagine yourself confronted at the supermarket shelf with two cans of ravioli, of which one is red, yellow, green, blue, and orange with homey scenes depicting the exquisite pleasures of Italian haute cuisine, and another which has minimum labeling giving only hard facts about ravioli in black and white, you can appreciate the relative appeal of the first. To oversimplify, part of the purchase price allows the retailer and the manufacturer to employ yet more advertising in the mass media, to induce in the consumer and others a fervid love of the brand name, so that he or she will continue to buy the ravioli—provided it tastes good enough when actually eaten. Ironically enough, a successful promotional effort can create a sufficient demand and therefore overall profit so that the fancy-can ravioli can be sold at the same or even less cost to the consumer than the plain-can ravioli, whose manufacturer probably goes out business.*

Nevertheless, the consumer pays to be persuaded to buy. What are the options available? First, fully conscious that we are paying to be persuaded and for disposal, we can continue as before. Alternatively, the consumer can attempt to cancel the effects of his or her previous gullibility by resisting packaging and advertising appeal and being obdurate in purchasing the product and not the package. The consumer can deliberately ignore the blandishments of advertising and when contemplating purchase concentrate on the salient features: ingredients, net weight, easily disposable packaging, and, after purchase, on the acid test, quality. Businesses thrive on supplying what customers demand. If the consumer demands more or better ravioli and less-fancy packing for the same price, then that is what the manufacturer will provide.

The on-going process of persuasion to waste, wasting, working to acquire the spending power to pay for waste and its disposal and the persuasion to do so, then wasting more, and so on, in a vicious circle, can be stopped only at the point of purchase—and only by the individual. The expression used in marketing to designate the power of the individual to get what

* A recent and very welcome innovation is the unbranded packaging of many consumer products.

he or she wants is "consumer sovereignty." If we allow ourselves to be manipulated like puppets we have abdicated that right and have put the power into the control of the advertisers. The customer is always right, because the seller exists, as such, only to satisfy the customer's needs.

The individual is powerful, not only in purchasing power in relation to the operation of the market, but also as a member of society in the influence he or she exerts on other people. A society, indeed, consists of, and is created by, a community of people in interaction.

Relatively few individuals have no influence on anyone else. Although the main emphasis of RICH is not missionary in nature, its gospel can best be spread throughout a society by the example of the few who have been aware of the inefficiency of present consumption habits.

Individuals can reduce the proportion of waste in the throughput process in three distinct ways:

1. *By Selective Buying*. The influence of individuals exercising consumer sovereignty reaches back around the throughput cycle to shape the production process. If the customer refuses to buy shoddy, badly constructed goods and leaves them on the shelves, quality and durability will quickly become the watchwords of the retailer, the wholesaler, and the manufacturer. If goods which are energy wasters or are impossible to repair when damaged or worn are rejected outright—before purchase—nobody will make them. Selective buying simply shifts the problems of waste back from the customer to the producer, and finally annihilates those problems. Undoubtedly there is a price to be paid. Energy-efficient appliances, for example, may demand more sophisticated design, skilled workmanship, scarcer materials, and thus could cost more at the retail outlet. The customer, however, gains in energy savings, reduced pollution (even if it is only noise from a cheap refrigerator), and the time and energy otherwise involved in repairs or disposal and re-

placement. The purchaser must also invest time in selective buying and consult consumer reports which set out clearly the expected performance, energy efficiency, guarantees, and ease of repair of goods currently on the market. (Incidentally, if these reports were more widely used they would become cheaper and, at the same time, more effective in reduction of waste.)

One of the psychological results of the general acceptance of the throw-away culture is diffidence on the part of the consumer about demanding any standard of performance from goods purchased. If we allow useless articles to take up space in attics, basements, and closets, we are depriving the retailer, wholesaler, and manufacturer of valuable information and encouraging more wasteful production. In a way the sensible desire to invest wisely over the long term is inseparable from the responsibility to reduce waste.

A less obvious aspect of selective buying is resisting the lure of fashion, particularly in clothing. If individuality, even eccentricity, in dress were encouraged rather than frowned upon, a triple benefit would result. Life would be visually more varied and stimulating, clothes would be made to last longer and to be aesthetically pleasing, and we would have more living space.

2. *By Efficient Use.* This is the most obvious and fruitful area for the Reform of Inefficient Consumption Habits. Although consumption takes place everywhere, the most generally applicable examples are in domestic consumption—and especially in nonluxury goods such as water, electricity, heating fuel, clothing, and gasoline.

Water

Profligacy with water can be attributed to a deep-seated but mistaken belief that water is in limitless supply. This is one of a set of paradoxically strange notions: the more it is held that water is abundant and should be used accordingly, the scarcer the water becomes. The belief, in time, self-negates. Even residents of water-rich areas should use no more than they need if only because channeling, piping, and purification cost money, energy, time, and labor.

Water is not destroyed in use; it is polluted. Only a very small proportion of an industrialized society's water requirement is for drinking or for watering crops; most of it is used for cleansing and for the production of hydroelectric power. In areas where water itself and/or the resources required to decontaminate it are scarce, the strategy would seem to be to use, and thus pollute, less of it in the first place. Is that possible?

"Cleanliness," it is said, "is next to godliness." If so, it comes a very poor second (except of course for the ungodly). There is no question, however, that North Americans have a preoccupation, if not an obsession, with being clean. A clue to the psychology of this phenomenon is found in the story of the newly immigrated European on his first trip "home" who proudly boasts, "We Americans shower twice or three times a day"—to which his father replies, "You must get yourselves very dirty."

Water waste usually occurs where more water than is necessary is being used: in showers where only a small proportion of the water actually washes the body; or in cleaning vegetables and fruits, which can actually be better accomplished by steeping briefly; or in car washing, which when done in the driveway with a bucket needs about a hundredth part of the water flow used in automatic car washes; or in lawn watering, which uses on the average about four times the amount that is good for the grass. These are examples of water-waste on the part of individuals but some industries are equally guilty—and on an even larger scale.

Electricity

Electricity, like water, is taken for granted. It is not in limitless supply. Indeed, if the current fears of running out of fossil fuels are well founded, electricity as a substitute may soon be required in much greater quantities for such purposes as home heating and transportation. When it is realized that it is just as convenient to conserve electricity as it is to use it, several simple practices are obviously indicated.

Turning off lights and television sets in empty rooms—and even in empty houses; the notion that burglary is prevented by leaving them on is so widely held that burglars too must be aware of it. Turning off ovens and burners on electric stoves when sufficient heat can be retained for long enough to accomplish cooking. Nutritionists are in agreement that we overcook vegetables to the point of losing substantial food values. Filling electric kettles only to the level immediately required. Using air-conditioners when and to the extent that they are necessary for comfort. Realizing that the purchase, transportation, storage, and repair of such gadgets as electric toothbrushes, back scratchers, massagers, and can openers take more money, time, and trouble than they are worth.

Heating and Cooling Fuel

The highest air-conditioning load in business skyscrapers in Calgary, Alberta, is in the middle of January, not in the summer, in spite of the fact that outside winter temperatures are close to zero Fahrenheit. The reason: faulty design which allows a huge expanse of glass to absorb the strong winter sun, which makes these buildings unbearably hot. It is interesting to note that the same idea properly applied could be very conservationist, since it should be possible to heat these buildings with solar power.

So much has already been said about petroleum scarcity that we need only list here the obvious measures that can be taken to reduce waste. Turning down thermostats (from 75–80 degrees Fahrenheit, which is both unhealthy and uncomfortable, to

68–70 degrees). Turning down the temperature level attained by water heaters: 120 degrees is hot enough for any domestic purpose, 150 is too hot. Shutting off heat in rooms rarely occupied. Installation of double windows, efficient sealing, insulation and subsidiary solar collectors. (Insulation alone achieves about 15 percent fuel and therefore money savings to the home owner.) Opening drapes on large windows (which themselves lose heat) to allow the sun to participate in heating and lighting.

Gasoline

When a 120-pound housewife takes her 3500-pound car to go to the grocery store to buy a pack of cigarettes, the energy efficiency of the trip, taking into account the weights involved, is slightly worse than that on the supersonic Concorde jet per passenger mile.

With at least one-fifth of national energy use being expended in private-car transportation, the need to reduce wasteful driving, where it exists, is obvious. Roughly speaking, there are two approaches to saving gasoline: drive less and drive more efficiently.

How can we drive less while still maintaining the automobile as a convenience, as a status symbol, an expression of machismo, as an extension of personality, or whatever? In other words, how can we reduce driving without changing our way of life? Presumably, there would be no loss involved in reducing the more irksome and unpleasant facets, such as are involved in rush hours, traffic jams, and difficult parking. The answers are obvious and require only a little careful planning.

As far as possible, avoid traveling in rush hours. Organize car pools for commuting and shopping trips. Patronize neighborhood stores and services. Use delivery agencies for such things as dairy products, newspapers, dry cleaning, groceries.

3. *Re-use and Careful Disposal.* From 1920 to 1970 garbage

production per person per day went up from perhaps ½ pound to about 5 pounds. The increase is continuing at the same or higher rate. Yet there is no necessary increase in happiness proportional to the rise in garbage production per person. Some of our more inefficient consumption habits can be reformed by using a little imagination. Almost anything made of wood, metal, glass, leather, plastic, cardboard, cloth, or paper can be adapted to some second use. There is no need for detail here, since most household magazines give detailed instructions on re-use of materials in handicrafts, furniture making, and so on. All we would suggest is a slightly different way of viewing used goods before they are discarded to become constituents of the growing garbage piles. If that sort of thing seems too much trouble, at least give away or sell apparently unusable items to those who can recycle them.

A much more significant reform can be instituted in the disposal of what might be called "habitual garbage"; of course, to the extent that selective buying and efficient use and re-use are adopted, the disposal of unintended by-products of consumption becomes less of a problem. The aim should be, first, to reform the present situation, set out in the table opposite which shows the main constituents of garbage as percentages of the total.

Careless buying habits reach around the throughput cycle to create the unintended by-products which in turn must be dealt with by using energy, time, and materials for their transportation and/or destruction. If we do not buy nonrefillable glass bottles we obviously do not have the problem of disposing of them. If we carefully plan our food buying we will not have to transfer the tainted items from the refrigerator to the garbage pail. Assuming even the most efficient "pregarbage" behavior, however, some problem of disposal will remain.

The most inconvenient, if not the most offensive, aspect of household garbage is its mixed nature. Almost every constituent of garbage is reusable or recyclable if it is isolated from the

Approximate Urban Residential Solid-Waste Generation*

CONTAINERS AND PACKAGING		37.9%
Glass containers (beer, soft drinks, wine, liquor)		10.9%
Steel cans (beer, soft drinks, food)		66.8%
Aluminum (beer, soft drinks)	less than	1 %
Paper, paperboard, and corrugated		14.6%
Plastics		4.2%
Wood Packaging		1.3%
NONDURABLE GOODS		29.4%
Newspapers, books, and magazines, office paper, paper tissues, paper plates and cups, other nonpackaging paper, clothing and footwear, miscellaneous		
FOOD WASTE		21.3%
YARD WASTE (variable with season)		5.1%
MISCELLANEOUS INORGANIC		6.3%

*Source: Solid Waste Management Branch, Environment Canada.

others; nonreturnable bottles can be collected at a glass-recycling depot; newspapers and clean paper packaging can be recycled as paper; organic materials (like food waste) can be used as garden compost; and so on. Within the CS_1 scenario, it is not recommended that precious time and heavy labor be devoted to recycling. The whole idea of CS_1 is a relatively effortless change of habits. As far as our discarding habits are concerned, the procedure is first to participate in (or encourage the establishment of) local recycling plants and then to follow up with household arrangements for separating glass from paper, organic materials from inorganic, as appropriate.

What is being proposed here is something more subtle and probably easier than a change of basic ideals. It is only a shift in perceptions. If we stop looking at garbage as something dirty to be taken out of sight as quickly as possible and consider it instead as "second-cycle material," in most instances having a positive rather than negative value, we will come closer to the truth of the matter.

RICH must, as the very name implies, make us richer in a subtler sense than does sheer accumulation of unwanted, unneeded things. In fact, to borrow from a TV ad frequently aired in 1977, the money and energy saved from RICH can now be used for whatever we consider the "finer things in life." This is the spirit and philosophy of doing more with less.

A Letter for Sammy from Lenny Lease

"Buyer's Regret." The two words found an immediate and poignant echo in Lenny. He was astonished to find that the old familiar feeling had a technical name. His marketing professor was pompously describing the use of advertising to reduce the feelings of remorse which often assail consumers immediately after impulsive buying. The danger, he said, was that the regret would stay and inhibit future purchases of the same kind of goods. TV commercials, newspaper ads, and so on, therefore had a dual purpose, not only to encourage new spending but also to induce feelings of contentment and pride of ownership about goods already purchased.

Lenny well remembered how in his poor childhood home in the Bronx such a possession as an electric organ had, at first, given his family a sense of richness and excitement. But the euphoria never lasted for long. Within a week, when the whole family had played through their repertoire several times, the organ had become an accusation: sitting there, taking up space, silently screaming its purchase price at everybody in the Lease household, who would all proceed to blame one another for getting the crazy idea in the first place. The TV image of the happy family clustering around singing their hearts out, night after night, faded quickly. Everybody contrived not to look at the offending organ—everybody except Lenny, who eyed it sideways, figuring out how many superburgers, how many bikes, how many pairs of Levi's they could buy with the money spent on it.

As the professor was explaining so detachedly, the same thing happened again and again. Anticipation, purchase, euphoria, letdown, regret, reassurance ... anticipation for something new. Lenny remembered that, as the financial situation in the Lease household had slowly improved, the regret shifted from money spent to space being taken up and, for Lenny at least, a feeling of being taken.

Lenny had been smart enough to avoid the draft and too smart to avoid college. In his first year, he hung around admiringly on the fringes of the Sammy Squander crowd. Lenny saw in them, particularly in Sammy himself, a certain touching, middle-class innocence. He soon realized, however, that they had the same problems as himself, only on a grander scale, which he couldn't keep up with. Lenny soon gravitated to the company of Angus McThrift, if only because the Canadian never did anything Lenny couldn't afford. With typical street urchin's perceptiveness, Lenny could see that, of the whole crowd, Angus was the one who would "make it" in life. He listened carefully to Angus's quiet philosophizing about "common sense" and "wise disposition of resources."

It was just after the "Buyer's Regret" lecture that the idea of renting and sharing crystallized in Lenny's mind. He began to see the possibilities of renting things for his own use, but more important than that, the tremendous scope for rentals as a business. He studied the problems of seasonal commodities, load equalization, durability, and obsolescence in consumer goods and became the class expert in all aspects of shared use. Soon he was putting his theories into practice and, by the last year of college, he was renting out typewriters, skis, bicycles, coffee urns, season tickets for the ball games. He collected commissions for subletting apartments during the summer and even for the shadier business of sharing term papers.

It was as the entrepreneur of temporary need satisfaction that Sammy Squander remembered Lenny, as he tore open his old friend's letter.

Castel-Del-Rey
Laguna Beach
California
July 25th

Hi Man, Sammy,

Long time no see—mainly because you don't get off your butt and get over here. It's real, baby—and it's far out, like you wouldn't believe it. The bread is good, forty thou a year but I don't even spend it all 'cause I buy nothing. No, man, not me. That was my trip years ago. Now, Sammy boy, I rent. Seriously, Sammy, renting gives me the freedom to move (and, boy, am I on the move) and at the same time lets me trip out in almost any direction. I've spent the life of a millionaire with just peanuts. To begin with the condominium at Castel-Del-Rey.

On the first floor is a gigantic boutique, where clothes are chosen on Monday for the week. A fleet of cars awaits in the garage and the aesthetic coquette can match her chosen car to the color of her dress rather than vice versa. Books on all subjects exist in the library on the second floor and a record-lending library exists on the third. Color television is rented for the rainy nights, while in midsummer the residents may either perfect their tennis on the communal tennis courts, learn aquatic sports from a variety of skilled instructors, or sail a yacht for the weekend (the waiting list for this last item is very long).

A family on the seventh floor has rented an authentic Picasso lithograph for six months, the girls in Apt. 56C have a living room full of rented soapstone carvings and a complete set of genuine Navaho ritual jewelry. The young executives in 715A are tripping out in a Louis XV salon.

Well, Sammy boy, there's action here whenever you want it. Can't help but feel that if more people did the whole rental trip there'd be more to pass around, less to produce, and less to pollute. What I get most out of it, Sammy boy, is change and all the kicks that come with change. Why get stuck with all that junk anyway? Get sick of my things third time I wear them. Get the blues driving the same car. No, man, in this world of change might as well move with the wind. The cure for transience may be more transience. Own nothing and you won't get the bum trip you guys have every time you love it or waste it.

Be seeing you,

Lenny

9. SHARING BY RENTING

One way to conserve is by reducing the need for further throughput. We can reduce our demand for new production either by changing our tastes (which implies value change) or by making better use of existing production, that is, sharing what we now have. The rental scheme, based on simple ideas, is a strategy for conservation which possesses many desirable side benefits.

1. *Renting reduces the demand for new products by promoting load equalization in the use of existing products.* The rental scheme as a strategy for conservation is principally based on the fact that *we own many more things than we can use in a given time period.* Because we own these things we remove them from the public domain and prevent others from using them. If a satisfactory method of sharing what we have could be devised, then more people would enjoy the luxuries of life without the need to produce ever more quantities of redundant commodities. Renting, which really means possessing only when using, is one method of achieving this greater efficiency. Hence a rental society will reduce production throughput by making consumption more efficient.

Assume we have unlimited energy, unlimited raw materials, zero population, a true land of plenty; would we have perpetual exponential economic growth in consumer goods? No. There is one insuperable limit to the growth of consumerism—time. The absorptive capacity of individuals is constrained by the simple fact that there are only twenty-four hours in a day.

As time progresses we acquire more and more possessions. However, there is a minimum amount of time which must be employed in enjoyment of each of these items. In fact, the more possessions we have, and the more complex they are, the greater is the total enjoyment time needed. A new suit, a car, a finished basement, a stereo sound system—all take progressively longer to enjoy properly.

In spite of this fact, we accumulate possessions at a very steep rate. We all come to an ideal point where we have an optimum time per possession and just the right number of objects. Any more possessions is pure waste. This "ownership waste" is most prevalent in our society. At any one time there are masses of unused goods in our possession. This includes cars parked in the driveway, clothes hanging in the closet, unread books on the shelves, unplayed records, empty summer homes in winter, empty winter homes in summer. This accumulating redundancy is aided and abetted by advertising campaigns which attempt to stress the so-called joys of ownership, even when the goods are not enjoyed.

All of us could, through some private introspection, arrive at our own conclusion as to the extent of our own ownership waste by noting how much time we actually spend with each of our possessions. As children we were sometimes told by our parents that if we did not play with the various toys we had accumulated over Christmases and birthdays they would be given to other children. If we were to apply the same principle as adults we would have something like a rental society.

Technically speaking, the whole rental scheme can be related to the question of the "load factor," the measure of use of a piece of equipment. It is possible to prove that in most situations load equalization improves overall efficiency of the system, In this particular instance we save by needing less.

An example about the perennial American toy, the automobile, is instructive. In 1977 there were well over 100 million vehicles in North America for a population of 230 million. In spite of that, 8 or 9 million cars are produced every year with all the accompanying throughput cost. If we possessed a general rent-a-car system we would well manage with the stock of 100 million vehicles. Those of us who walk to work and use the car only on weekends would rent the car for the weekend. On the contrary those who use cars on weekends only and play golf or watch TV on weekends would rent them on weekdays. There would be no loss of satisfaction yet a substantial conservationist effect would result.

2. *Renting is compatible with the production of durable commodities.* Obviously, rented things must be durable. We certainly cannot rent an ice-cream cone or a steak. The more durable an item, the more it is suitable for this purpose. Although renting itself will not *cause* products to be durable, it will create an incentive for producers to make them last. Witness for instance the greater durability of commercial-grade carpeting found in hotels as compared with the more fragile items sold for our personal use. Witness also the telephone, which we all have in our homes and which was built to be rented; it is so durable that it is almost an embarrassment to its manufacturers, who cannot easily introduce new improved models. It can take use and abuse with impunity. Note also the vehicles that are specifically built to be taxis. These vehicles can absorb 100,000 miles with acceptable wear and tear as compared with the cars that are normally offered for sale and which self-destruct in a very few years.

This creates an interesting paradox. Whereas renting would be an incentive for durability on the producer's or owner's side, it could become an excuse for abuse on the consumer's. We have all experienced the syndrome of "what the heck, this thing is rented so why should I take care of it?" In spite of that tendency, there is an easy answer: prevent abuse by the consumer by imposing penalties for improper use. In any private enjoyment of a public or semipublic good there are mandatory rules. There are dos and don'ts for a tennis club, a rented apartment, a rented rug shampooer. If the consumer decides to play the role of the vandal it is easy to impose fines to discourage or make up for mistreatment of the rented object. A careless apartment tenant has some accounting to do to the landlord at the end of the lease. If he has improperly used the premises and damaged them beyond normal wear and tear, he has to pay for repairs.

Durability is a design feature that must be built into the product. It is easier to monitor consumer behavior in a rental scheme to avoid improper use than to monitor producer behavior in a selling scheme *unless the producer is enjoined by law to guarantee the item for a certain period of time*. If, as is generally true, *caveat emptor* prevails, the consumer is left to fend for him- or herself. Many of the warranties that exist for various products tend to be more token than real, and if real are difficult to implement effectively.

3. *Renting is compatible with high quality.* Durability is not necessarily synonymous with quality but, nevertheless, often closely related to it. This relationship allows us to conjure up an image of a possible rental society with very high-quality products in shared use among many consumers. This image is in fact confirmed by an examination of what should be the *least* expensive car to rent.

In a reasonable rental system there are three elements making up the rate: the depreciation cost; the interest on the capital tied up; and the profit accruing to the entrepreneur. If we assume a reasonable rate of profit and if we assume the monetary interest on the capital to be determined by general economic conditions, the principal and most significant cost element in renting becomes depreciation. The higher the depreciation, the higher the cost, and vice versa.

Applying this analysis to automobiles, we come up with the

profile of the ideal rented car. It is not a Ford Pinto or Chevrolet Vega. It is not even a Volkswagen. It is more probably a Mercedes-Benz, a Rolls-Royce or a Volvo—cars that are known for low depreciation; in fact, the Rolls-Royce appreciates in value over time.

Why then, it might be asked, are these cars not rented at low rates? The answer is twofold: first and foremost, luxury-car dealers face an inelastic demand and therefore can charge very high prices. If a Rolls-Royce salesman rents the Rolls-Royce for a year at high rates, and then benefits from an appreciation of the value of the car, he is not likely to return the extra money to the consumer. He will make as much as he can. However, if he were content with a smaller profit, the cost to the consumer would be substantially lowered and we could reasonably rent Rolls-Royces.

The second reason pertains to the legitimate apprehension of vendors when faced with the potential irresponsibility of some consumers. A Rolls-Royce may be durable given normal use, but it can be wrecked if used on drag strips or in demolition derbies. Once again we face the problem of policing the consumer, which is a difficult yet surmountable problem. The general idea remains intact, however: renting to responsible consumers is compatible with high product quality.

4. *Renting allows consumers to enjoy a higher standard of living than is possible through ownership.* Many people who cannot afford to buy a Mercedes could probably afford to rent one for short periods of actual use. If we generalize this we can again conjure up an image of a consumer (à la Lenny Lease) who lives very well without missing anything: a rented Mercedes, a rented Picasso, a rented summer house. . . . Of course, it is easy to exaggerate and produce a nightmare vision of impermanence and transience. But the plain fact remains that the consumer would be empowered to improve his or her standard of living by renting.

There is evidence to suggest that for dozens of consumer goods brief enjoyment is more than enough. Economics teaches us that the law of diminishing marginal utility is probably universal. Psychology indicates that certain needs can be satiated. Ordinary common sense tells us that there is such a thing as freedom from desire. The exhilarating experience of a sports car on a mountain road, the opulent luxury of the royal suite in a posh hotel, the convenience of the private swimming pool—all meet the ceiling of satiety. Once that is reached, the owned good becomes redundant, whereas the rented good is returned to the public domain.

Therefore, if it is in fact true that continued possession and use of most products dulls the glitter and glamour associated with the initial enjoyment of them, renting would allow us to possess the goods for the period of maximum psychological enjoyment and then return them for the next user.

5. *Renting promotes greater diversity in life-styles and modes of enjoyment.* An owned good becomes a bond, a weight, a reminder of previous times. A fifty-year car that is bought is going to haunt its buyer for fifty years. The scenario of durability with ownership may mean an excessively dull existence with two or three twenty-year suits, six ten-year shirts, indestructible furniture, etc., etc. Durability, which is a virtue in itself, becomes a crashing bore.

To obviate this problem, the rental scheme allows diversity. With renting it is possible to change cars not every two years as many do, but every weekend. The permutations and combinations boggle the mind, and the confirmed Big Rock Candy Mountain climber finds, indeed, that he can do more with less—by owning very little and renting everything.

As a passing thought, a smorgasbord—the very image of plenty available to all the guests at the same time—is actually more economical than any attempt to recreate the plethora of food on an individual basis. Renting is, in this sense, akin to the smorgasbord: economical, diverse, and, if applied intelligently and responsibly, conservationist.

6. *Renting is egalitarian.* A public good is by definition available to the public. When the Palace of Versailles, the British Museum, the Sistine Chapel, Disneyland were opened to the public, a mini rental society came into being. By buying an admission ticket the average citizen could bask in the glory of an experience of beauty and joy that had hitherto been reserved to an elite. The proliferation of public, semipublic, and rented goods contributes toward developing a less class-conscious society. When renting is available to almost everybody, the prestige effects associated with sole ownership (itself used as a method for social stratification) become lost. Who can impress his neighbor with a Cadillac if that same car is available for low-cost rental? It follows, therefore, that a rental society would be egalitarian and reduce both the real and the psychological income inequalities between people—one of the avowed goals of the American, the Canadian, and almost every other government in the world.

The object of acquiring goods is to enjoy them. In the strict legal sense of Roman law the three characteristics of property rights were *usus* (the right to use); *fructus* (the right to enjoy the fruits of); and *abusus* (the right to alienate, abuse, sell, or destroy). Renting in a sense conveys the first two, a sort of a usufruct with abuse missing.

The philosophy of renting is based on two ethical principles: experience and stewardship. We are here to experience happiness, not to buy it. We do not own our wives, our husbands, our children, our parents, or even the fleeting unforgiving moment. We experience them. We cannot take our worldly possessions to the grave either. We are but stewards of the earth's resources and it behooves us to husband these appropriately. Stewardship is central to the very idea of a conserver society, emphasizing harmony with nature, a balanced ecosystem, a peaceful environment.

But who should actually own goods in a rental society? Obviously above and beyond nature's bounty, there is the surrogate world of shoes and ships and sealing wax which must belong to someone. Renting is payment-for-use to the owner. Who is the owner?

There are four answers. First, it is possible to imagine ownership, the final sovereignty over goods, to be vested in society itself as represented by the state. That would be tantamount to socialism or utopian communism, which, in some senses, are both similar to rental societies. Such a device is heavy with ideological presuppositions which in 1977 are not those of mainstream North America. The very mention of "creeping socialism" frightens a great many people, who, rightly or wrongly, view it as a threat.

The second rental option is the corporate one with ownership vested in large rental corporations, which would lease their goods to the general public. This would imply a proliferation of the Hertz-Avis-United-Rent-All framework and would possess both advantages and disadvantages. The advantages of efficiency would be counterbalanced by excessive economic power concentrated in a few major corporations, which could, in effect, hold the rest of society hostage.

The third variant is a consumer-cooperative one, with the users actually owning the goods collectively. The model for this is a condominium, a private club, or an agricultural machinery cooperative. This variant allows for decentralization, grassroots control, and smaller-sized pools of consumer goods.

An ideal rental system would probably opt for a fourth way, which would be a judicious mixing and matching of the first three. Elements of the statist, corporate, and cooperative variants would combine in an optional mix. Private ownership of certain goods must persist for obvious reasons. We surely are not going to rent toothbrushes and underwear!

The rental-based conserver society would be called upon to assess the degree to which goods are of a "public" character. At one extreme there could be fully public goods, such as superhighways and bridges and, at the other, fully private, such as the toothbrushes and underwear. In the middle could be a large pool of semipublic and semiprivate goods shared in rental systems varying from the corporate to the cooperative options.

Renting is not without disadvantages. There is the question of the lease period. Renting over too-long periods is equivalent to ownership, since the commodity would be removed from public use without necessarily being fully enjoyed by the lease. It is possible to lease a car and leave it idle in the garage, resulting in no conservationist effect.

There is also the double-edged sword of load equalization. On the one hand utilization of idle resources is efficient and to be recommended. On the other hand if, through rental schemes, all cars in circulation were to be on the road at the same time it would create a nightmare of traffic jams, pollution, and energy use.

A third drawback to a pure rental scheme is inconvenience. People who are members of a tennis club and who have to sandwich their tennis between 4:45 and 5:30 on Wednesday or have to wait to use the exercise equipment in a gym, all have experienced the inconvenience of renting rather than owning. It is, of course, more convenient to have a private tennis court or a private gym. The question is, does the removal of the inconvenience justify the tremendous waste represented by a private tennis court? Or can there be, instead, a way of minimizing the inconvenience by more careful load equalization? This points to a better management of time, discussed in the next chapter.

The final drawback is the psychological value attached to

ownership, which is lost in the rental society. To rent a Rolls-Royce is not the same as owning one, in the mind of the true aficionado. The prestige may be lost and also the little irrational joys of ownership, such as spending the whole Sunday afternoon washing the hubcaps, waxing the right fender, and polishing to a spit the Rolls-Royce insignia. Less subjectively too, there is the question of true or alleged health hazard associated with renting. The idea of sharing clothes is repugnant to many (except actors, who share costumes as part of their job), and if a person were to be asked, without preamble, to rent his or her bed sheets rather than own them he or she would probably react in horror. Yet this is exactly what we do when we stay in hotel rooms. In some situations the objection is legitimate, in others a mere cultural whim. A feasibility study should take account of both.

To some extent the rental society is already emerging in North America. Some notable trends can be highlighted.

Car leasing rather than buying is becoming increasingly popular. A significant proportion of cars in Canada and the United States are now being leased, and almost every dealership offers the "lease or buy" option. Two possibilities are usually offered, an open-end or a closed-end lease. The former is a sort of conditional sale with lease payments culminating in an option to buy the car after two or three years. The closed-end version is a straight lease as of an apartment. The attractiveness of leasing to the lessee is twofold: it avoids tying up capital and there are tax advantages when the leased car is used for business purposes. If car leasing were done with shorter lease periods to achieve load equalization, and if cars were made durable to be re-leased after three years rather than released to the junkyard, the conservationist effect would be high. The consumers have shown their willingness in this respect. It is now up to the producers to manufacture more durable cars and to shorten the lease periods.

Office equipment is increasingly leased. In an article in the

Elements of the Corporate Rental Model Circa 1978

The following commodities are among a wide range of items currently rented in Canada. (This is an illustrative and not an exhaustive or exclusive list.)

Item	Functional Items	Luxury Items	Communications and/or Transportation Commodities	Commodities Involving High Throughput in Production	Commodities Involving High Throughput in Uses
Apartments	X	X		X	
Furniture	X	X		X	
Household appliances	X	X		X	X
TV sets	?	X	X	X	X
Radios	X		X	X	X
Air conditioner		X			X
Carpet shampooers	X				
Lawn mowers	X				
Costumes (formal)		X			
Cleaning services	X				
Catering services	X				
Restaurant services	X	X			
Telephone receivers	X	X	X		X
Computer time	X	X	X	X	X
Airplane seats			X		X
Ship passage	X		X		
Train/bus	X		X	X	
Rental Automobiles		X	X	X	X
Chauffeur-driven taxis		X	X	X	X
Motel beds	X		X		
*Money	X				

*For example, when a loan is taken and interest paid for temporary use of the money.

Financial Post (March 27, 1976) titled "Should you lease that desk?" a good case is made for leasing business equpiment. Exactly the same analysis applies here as with car leasing. The producers should manufacture durable items and shorten lease terms. The incentives for the consumers are, again, the enjoyment of the commodity without tying up capital and, probably, a tax advantage.

Household appliances, computer time, televisions, radios, and even art are all entering the lease market.

The residential sector is another eminent area where leasing could replace ownership. The one-family dwelling is a great ecological waste even though it contributes to the quality of life. To rent an apartment in the present corporate rental model is not, however, a satisfactory answer, since what it does is build up the profits of the landlord (in the absence of rent control) and deplete the savings of the tenant. The answer may well be the condominium dwelling.

There is no reason to suppose that a condominium need necessarily be a drab, uninteresting, regimented apartment building. It could be fashioned along the lines of Moshe Safdie's Habitat '67 or better Habitats for 1980 to 1990. It could be pleasant, stimulating community living with privacy also built in. Nor is it necessary to suppose that condominiums would have to exist in an urban setting. Rural community living is perfectly possible and, to some people, highly desirable.

10. THE MANAGEMENT OF TIME

This short chapter outlines the companion piece to the rental society: a better management of time. It is a potentially powerful conservationist technique that is eminently suited to CS_1 because it is effortless.

Renting is a form of sharing which assumes that everyone will not want the same goods at the same time. This is an essential condition for the success of such schemes. In a general way, it can be stated that an important source of waste is the tendency planners have to endow their operational systems with a capability to meet peak loads rather than just average loads. Yet peak loads occur a fraction of the time, and therefore at any other given instant there is enormous unused capacity. Some examples:

The City of Edmonton in Canada spends millions of dollars adding lanes to its superhighways to meet a demand that exists between 8:30 and 9:00 A.M. and between 5:00 and 5:30 P.M., i.e., the rush hours. The rest of the time, the superhighways are very slightly used. An alternative conservationist solution would be to eliminate the rush hour rather than catering to it. This can be done through staggered working hours or flexitime—in short, a better management of time.

Automobiles are endowed with horsepower, stereo sets with amplifying power, electrical systems with carrying power, etc., that are rarely used to full capacity. Speed limits effectively curtail the drag stripper and the aficionado of 110 mph average speeds—although many current-model cars still have awesome potential. The amplifying power of stereo sets is often limited by irate neighbors, and unless one has a sound-proof basement much of the wanted amplifying power lies unused. With electrical systems, generally, the situation is different because there are peak loads, especially from 5 to 8 P.M., when in apartment buildings the stoves, the refrigerators, the televisions, lights, and perhaps the washers are all on at the same time.

Some of this planning for peak load can be made unnecessary by an intelligent management of time, both individual and institutional.

Time may be regarded from an individual point of view as the ultimate nonrenewable resource. Although superficially renewable—there is this Tuesday, next Tuesday, and the one after that—Tuesday, June 28, 1977, is gone forever. The great decisions that affect the management of individual time fall into two categories: life-cycle choices and day-to-day choices.

Life-cycle choices relate to the different possible allocations of time between education, work life, and retirement, whereas day-to-day choices concern sleeping vs. waking hours, and within the waking hours are options between work, chores, active leisure, idleness.

In an ideal conserver society, individual time would be managed in a fashion that would simultaneously allow both a high quality of life and an economy of resources. The actual techniques to be suggested are too numerous and specific to describe here, but the reader is invited to appreciate the potential of a better management of individual time as yet another technique of "doing more with less."

Management of institutional time offers the most promising implications for conservation. Daylight saving time is one excellent example: this comparatively simple measure accounts every year for very substantial energy savings, and some have suggested that it should be kept all year round.

Other examples of better management of institutional time are:

Blurring the distinction between weekday and weekend. The weekday *vs.* the weekend distinction is an important source of waste. What it creates in effect is severe strains on certain systems that are forced to perform at peak loads (such as Friday-night traffic) while simultaneously other systems are underutilized (such as the downtown area on Sunday morning).

Staggered work hours. This idea is obvious. It implies not only flexitime but perhaps blurring of the distinction between night and day, introducing flexibility in the lunch-hour period, etc. Staggered vacations are more or less accepted in North America (except for Christmas, Easter, and the national holidays). In France, on the other hand, this concept is unthinkable and Parisians take vacations only in August. The religious fervor associated with the indispensability of an August holiday creates an imbalance noted by tourists: Paris is nearly always empty in August and the Riviera intolerably full.

Blurring the work/leisure dichotomy. "Work" and "leisure" are subjective perceptions. A more intelligent allocation of time between so-called "work" and so-called "leisure" could alleviate problems associated with unemployment and lead to greater work effeciency at the same time (see Chapter 20 on the squander society).

Fees for piecework vs. *salaries for time spent.* Studies have shown that remuneration based on time spent (which is the idea behind a wage or salary) is not particularly conservationist. The worker has little incentive to get things done. The typical plumber, electrician, or repairman making a house call, the consultant paid on a daily basis, all adopt a taxi-meter mentality and have every incentive in the world to linger and work slowly. By contrast, piecework is an incentive to get things done. It is therefore more conservationist, although its general acceptability in 1977 must remain a question mark. We are all slaves to a system of trying to minimize our individual shares of producing the pie, not realizing that this attitude may actually make the whole pie smaller.

Overall, the management of time lies at the root of the very notion of conservation, which is nothing more than a question of time preference: if we choose the present, we consume; if we opt for the future, we conserve. The management of time becomes tantamount to the management of conservation.

11. CONSERVER TECHNOLOGY

A conserver technology would have one humble, but still ambitious, aim: it would do more of what we want done with less scarce physical resources than are used now; it would maintain or even improve our material standard of living but create less waste in the process.

A conserver technology falls far short of an ideal. The ideal would be a perfect interaction between the biological system of the human being and the ecological system of its environment and therefore would be at least as sophisticated as both; there would be no waste, no pollution, no unintended by-products. Indeed, there would be no by-products at all since the processes of production and consumption would be so well thought out that all the materials and energy emanating from them would appear exactly where and when they would be most required or desired. Such a technology would use natural resources only at the rate of nature's replenishment and in accordance with foreseen needs. It would make use of only the kinds and amount of human energy that people want to expend. It would even guarantee happiness in the individual, harmony among men and women, and peace among nations. This, of course is science fiction.

At a lower and more realistic level there is possible a truly efficient technology using scientific knowledge to its fullest. The major barriers are institutional and value based. Nations do not share information freely, companies do not communicate, individual scientists do not always share their discoveries. Research itself, whether it is conducted at the individual, corporate, or government level is very often primarily directed, not at the increase of understanding, but at gain in prestige, profits, and power.

If we accept the idea that it is impossible and perhaps even undesirable to change the basically competitive nature of our society, can technology be significantly improved? The scope within the limits discussed above is considerable. During the recent era of apparently abundant resources, the tendency has been to use the fastest, cheapest, easiest production processes, without forethought for the environment and our health. Consumer demand has been for flashy, throw-away goods. Research has been concentrated on making sales.

It is a tall order indeed to reform the technology which has supported the affluence and the effluence. Contrary to popular supposition, a conserver technology would not be primitive; it would not mean regression to the oxen plough and the hand loom. In fact, a conserver technology, at least within the CS_1 scenario, would be highly sophisticated, dynamic, and imaginative. As exemplified by the miniaturization of computers, small may be beautiful but not necessarily wooden. It takes more scientific and technological expertise to avoid waste than it does to create it; and this is especially true if we continue to desire comfort and convenience on a large scale while at the same time avoiding arduous and repetitive heavy labor.

The aims of a feasible conserver technology would be to substitute relatively benign and abundant resources for scarce and polluting types we now use in our production processes; to improve the durability and performance of the products themselves; and to make recycling a real economic possibility. These are all design problems, the solutions to which are certainly within the scope of present knowledge.

It probably need not be mentioned again that clean renewable forms of energy should largely be substituted for fossil fuels, but we must be realistic. There simply are not enough rivers and waterfalls which can be dammed to create hydroelectricity for all manufacturing processes. "Soft" energy, from the sun and the sea, does not produce sufficiently high levels of concentrated power. Where fossil fuels will still be required we must use those sources which are accessible, and by means of clean and efficient extraction methods. There are revolutionary proposals, for example, for the extraction of oil and coal underground, where the by-products, instead of becoming atmospheric pollutants, actually power the extraction process, once it has begun.

Turning a pollutant into an intended product is the neatest and most aesthetically appealing way of conserving. A good example has occurred recently in Canada, where a by-product of asbestos manufacture may become part of the process's regular output. Pre-concentrated magnetite, derived from the asbestos waste, is formed into pellets of an iron and nickel alloy, which can be used in the manufacture of steel.

The greatest challenge for a conserver technology is improvement of the performance and durability of consumer goods. "Performance," in the context of a discussion of conservation, can be measured by energy efficiency. The recent controversy over differing state governments' standards for refrigerators is a good example. If one manufacturer could produce appliances to meet the higher California standards, then surely so could the others. Generally speaking, we can assume that appliances which produce a great deal of noise and unnecessary heat are wasting energy and could be better designed.

The other aspect of the quality of consumer durables is, obviously enough, durability, but this concept itself is not so simple as it sounds. A laudable aim for conserver technology would be to increase the average useful life of products, but this does not mean simply making all the materials in a machine tougher. The complex nature of modern technological devices is such that hard metals rub against softer metals, both come into contact with plastics and rubber, and so on. The machine does not decay in a uniform fashion.

The design problem is to ensure that replacement of worn parts will be simple; the marketing problem is to make the parts available. It should not be a mere matter of chance if the soft parts of a machine are on the outside and easily removable; it should be a design requirement. Generally, products should be made to be durable, and failing that, they should be repairable. The producers' confidence in the dependability of goods should be reflected in warranties that are not designed, as they often are at present, to run out immediately before the machine breaks down.

So durability in manufacture means either better materials or easily replaceable parts or both. Infinite durability is certainly not possible and is probably not desirable. consider the implications of old-fashioned machinery enduring indefinitely while scientific knowledge and technological invention create potential replacements vastly more efficient and aesthetically pleasing. One of the reasons why Great Britain, the birthplace of the Industrial Revolution, is not competitive in manufactured goods is, paradoxically, that it is saddled with outdated but seemingly infinitely durable machinery. The related design problem here is that, beyond its useful life, a technological device should be "recyclable"; its parts should be readily separable to be adapted to other purposes or to be broken down to constituent materials, also to be reused. Cars, for example, could be put together in such a way that the body, the engine, the wheels, and the upholstery are separately recoverable.

The unsightly mess created by abandoned automobiles is one instance of the essential correspondence between the absence of recycling and the presence of pollution. A great deal of "post-consumer waste" can be avoided in the design of products and in the creation of "reverse channels of distribution." This means simply designing transportation and manufacturing processes so that discarded materials can be reused. Reuse of materials can substitute for the extraction of virgin resources—which is the point at which we began this examination. Conserver technology potentially meets challenges all around the throughput cycle: in the production-consumption continuum, in substituting recycling for pollution, in reuse of materials for further production, and so on.

Related aims for a conserver technology exist in architecture, in the design of industrial and urban complexes, and in distribution patterns. A more rational use of space and time should be built into all quasi-permanent installations, if only to facilitate the more rational of materials and energy. The movement of goods and people, which accounts for at least one-third of our energy use, can be replaced, in many instances, by the movement of information.

Most of the design problems are relatively simple and have been neglected because of apparent abundance of resources, along with a mania for immediate, easy profits. Others are real scientific challenges demanding new understanding of how materials break down, how they can be combined, how chemical simplicity or complexity is related to renewability and pollution. Indeed, the further improvement of a conserver technology depends on developing a scientific theory of conservation. Meantime, there is ample room for improvement of the technological situation.

12. FULL-COST PRICING

In order to stimulate the growth of recycling and of resource conservationist technologies generally, it is necessary to have an accurate idea of the "full costs" involved in production and consumption. There is a vested interest today in waste. Moreover, the costs of pollution and environmental degradation are frequently deferred to the future. Because many such costs are not accounted for in the prices consumers pay directly for goods and services, we live—or at least until very recently we lived—with the illusion of new-materials abundance on the one hand, and "nonprofitable" recycling industries on the other.

The idea behind full-cost pricing (FCP) is simple: to charge consumers the full costs of production, including those of resource depletion, pollution, and environmental degradation, in the purchase prices they pay for goods and services. How this might be accomplished in practice is not at all simple.

In 1975 Canadians spent just over $3 billion on packaging and related products. It is estimated that $4.2 billion will be so spent in 1978, an increase of 30 percent in three years. Because of its larger population, the respective amount for the United States is 20 to 25 times as great as that for Canada.

As we all know, much of this packaging ends up as garbage in one form or another. However, Canadians and Americans—who already have the dubious distinction of creating about six pounds of garbage per person per day—still seem to be prepared to spend huge sums of money to create more garbage. We are very much, as Alvin Toffler has asserted, into the "disposable society."

There are a variety of costs associated with the creation of garbage, many of which are not accounted for. The most obvious are the initial fees imposed upon us when we pay our municipal taxes for garbage removal and disposal. Beyond these direct costs there are a variety of indirect ones. For example, rarely do companies consider the "after the shelf" life of their products. Once goods have been sold from supermarket shelves, that is the end of the product as far as the supplier corporation is concerned. The fact that overpackaging can lead to immense problems in garbage collection and disposal is often forgotten. This is somebody else's problem. It is a "production cost," in other words, which is "externalized" to others now, or to future generations. Corporations would argue that through the business taxes they pay they are also bearing some of the burden of garbage disposal. This is undoubtedly true. But if the "convenience products" were not so well packaged it might be possible for the associated costs to be lowered considerably. The trend in recent decades, however, has been to more and more product packaging, the reverse of an environmentally appropriate design for future living.

There is a further category of costs associated with "garbage" creation which should be noted: the variety of "benefits forgone" from *not* using the valuable minerals and energy constituent in refuse. Garbage goes through many processes before it becomes "solid waste" (manufacturing, distribution, marketing). Therefore, it contains a larger per-unit energy investment than almost any other kind of "waste." We do not recapture this "energy" investment at present. We could, however, if the energy and materials constituent in solid wastes were recycled.

Similar situations exist as far as postproduction residuals are concerned. In recent years, the asbestos fiber content of tailings produced (and discarded) in the 1930s has been found to exceed the grade of new ore now mined. Consequently, these tailings are put through the concentration process again. This type of residuals utilization is usually referred to as "secondary exploitation." It is, however, recycling within any meaningful definition of this term, that is, converting "waste" into "useful" products.

Similar examples exist with resource depletion. In such situations, shortages may be price relative or absolute. In either instance significant problems may arise. For example, had the rate of discovery and development of oil and natural gas been the same in the 1960s as it was in previous decades, then the OPEC cartel would not likely have been so effective as it appears to have been in raising world oil and natural gas prices (and thus worsening the inflation of recent years). It could be argued that the excessive depletion* of oil reserves in North America during the 1960s has now become a *cost* that has been "externalized," that is, transferred into the 1970s.

Finally, there are a variety of costs associated with pollution avoidance which are not yet fully captured, but which, in any full-cost pricing system, certainly should be. For example, the energy invested in the purification of metal from which a machine is built is partly recoverable if the machine is salvaged after its useful production life is over (through recycling). This benefit is forgone if, instead of recycling the old, a new metal is mined and processed to serve in a *new* application.

Externalities are not only costs which specific firms do not bear at present but are also forgone benefits which are not realized under our present system of economic incentives and constraints. The benefits, although less visible than externalized costs, are of no less overall significance. The amount of energy savings which are potentially realizable through using old rather than new mineral resources are substantial.

What precisely these benefits are will depend directly on the technologies of primary processing and recycling. Hence, any

*That is, the cumulative consumption of oil, with respect to the overall development of indigenous oil, and of economic substitutes.

meaningful measurement of externalities must rest somehow finally and firmly on a given set of technological assumptions, conditions, and capabilities.

In assessing what costs are avoidable, the other available technologies, therefore, must always be considered. The extent to which given recycling methods recover basic factor (i.e., production) costs can thus become a key criterion in the classification of both primary and (the alternative) secondary or recycling technology.

It is unlikely that a society could be established in North America in which all "residuals" were recycled. There would always be leakages, in particular those energy losses resulting from entropy. Nevertheless, if producers and/or consumers were asked to pay the full costs of their production and/or consumption the economic system would be much more closed than it is at present.

The influence of government on all levels of business activity in any modern industrial state is pervasive. This influence is exerted by law, through regulation, through incentive taxation systems, or by subsidies, as well as by direct purchases of a variety of materials, goods, and services from industry, and through the ever-increasing employment which government provides, or the transfer payments it makes, to an ever-increasing population.

Heretofore, in North America, the impact of government on the pattern of industrial development has been largely perceived as a result of these types of direct activities. What has not been so readily recognized, until recent times at least, is the powerful indirect intervention which government may exert in industrial development. This may take the form of the conscious promotion of the development of one industry and, thus, the consequent but unconscious furtherance of the demise of another, often the competitor of the first. By supporting highway construction, but not railroad development, federal and state authorities undermine the competitive viability of the latter by giving a particular advantage to the former. Resource-conservationist technologies and recycling industries in the past have been similarly disadvantaged vis-à-vis resource-profligate technologies and primary resource users in North America.

Even today, there is very little conscious effort on the part of governments at most levels to promote resource-conservationist technologies or recycling industries. To illustrate. From the definition of taxation categories, in both Canada and the United States, depending on the specific technology employed, some recycling or resource recovery plants can be classified as engaged in a "manufacturing or processing function" or in "mining," whereas some others are neither, most likely fitting some obscure, ill-defined category.

The significance of this lies largely in the lost opportunity of recycling industries to qualify for existing tax incentives offered to manufacturing: accelerated depreciation periods for capital investments, as well as frequently lower than otherwise current income-tax rates. Equally, materials-recovery companies (that is, secondary materials producers or recyclers) frequently cannot claim the depletion allowances offered to their competitors, the primary resource extractive industries, such as mining enterprises. These two factors combine to seriously undercut the competitive ability of recycling industries to attract investment capital.

Intriguingly, although refuse (urban ore) is often richer in metal content than some of the natural ores mined, it is still not frequently recognized as an ore. This, plus the depletion allowances granted to mining companies, make the extraction of the natural ore a much more attractive financial proposition at present.

In addition to income-tax disbenefits, recycling companies—in Canada anyway—have often had to pay a 12 percent federal sales tax on the equipment purchased for processing purposes.

This tax does not apply to firms engaged in "manufacturing processing," but secondary metal dealers and processors—because of their ill-defined legal status—have frequently had to apply for exemption from this 12 percent tax every time they have acquired new equipment. Moreover, it is understood that this exemption has not been granted universally, as the meaning of the manufacturing-processing tax terms is not clear with respect to recycling. Indeed, interpretation varies frequently and, we can speculate, almost whimsically.

Hence, although there has been a generally acknowledged, strong government commitment to the development of primary industries, there does not appear to have been any appreciable government support—until recently—for those industries which *do* conserve resources.

In short, the impact of depletion allowances granted to mining enterprises (and not granted to resources-recovery companies) is punitive to recycling via the competitive-cost advantage given to the primary-resource producers. In a broader sense of capital allocation, recycling enterprises are not yet fully recognized as legitimate and vital industry, and thus are deprived of the taxation incentives needed to attract capital and talent for growth.

Government regulation plays a vital and, it seems, ever-increasing role in North American life. But is the exercise of even more regulative power by governments—at all levels—the sole answer to our problems of pollution, waste, resource depletion, and recycling?

Certainly if we go on the past record, we see—not always but often—that by the time regulations have been designed and then enforced, frequently through costly court cases, environmental battles may already have been lost. Moreover, under unrelenting pressure from industry to be allowed to "internalize" its benefits and "externalize" its costs of production, regulators are often forced to bend to the many arguments presented to them.

There is no doubt that government at all levels will have to set standards and police them and, equally, will have to clean up the no man's land of past pollution. Government should also support research into pollution abatement and appropriate technology since much of the gain here may benefit all industries, and the private sector is often unwilling or unable to bear the associated costs.

What about subsidies from government as a means to stimulate recycling and the growth of resource-conservationist technology? That is a very costly way to accomplish this end. Why? Because in order to get subsidies in the first place most firms have to do whatever is required of them in a particular, specified, way. With imagination there might be a much cheaper way of accomplishing the same ends. Moreover, industry—at present anyway—can use rivers, lakes, and so on to dump its wastes largely free of direct charge. To be really effective, subsidies would have to cover the full cost of pollution abatement before producers would accept them. These fees today are largely unknown. Moreover, how would such subsidy policies internalize costs to producers? Under the subsidy approach, costs would merely be shifted from one segment of the public, the pollutive or non-resource-conservationist industry, to the general taxpayer. Is this fair? Does it ensure effective resource allocation? A negative answer is called for on both counts.

What is really involved, in operational terms, is that technologies and approaches to production, distribution, marketing, retailing, and postconsumer waste management which are environmentally wanton will have to be punished, whereas those which are not will have to be rewarded.

The great strength of the market system is its technological creativity. The market system is the most efficient (least wasteful) manager of resources once the value of these resources is translated into monetary terms. This is particularly true in short-run periods. The market may exhibit weaknesses in its attempts to allocate resources over the long run, and this is precisely the area that is best suited for government intervention. Governments, although hampered by political considerations, can, in general, plan activities with a view to the longer term, but they are frequently less efficient than private enterprises in operating existing production, distribution, and retailing facilities.

In the light of this, full-cost pricing hinges upon a market mechanism which could incorporate externalities (that is, internalize them) into its decision processes. The internalization of costs could, of course, come about only as a result of the initial intervention of government. Thus, any proposed FCP system would be administered by the market but would, initially, have to be introduced by the state. As this developed, however, the role of the state would decline in direct proportion to the success with which solutions were developed by the market.

The most efficient intervention by the state into a market economy is attained via taxation. Selective and incentive taxation may alter costs, but it does not negate or impede the nature of market forces. Rather, by changing costs, it redirects these forces, ideally toward socially more satisfactory goals.

In addition, the choice to use taxation to bring about a closed-loop recycling-oriented conserver society, rather than to rely upon regulative, subsidy, or moral suasive techniques, is based on a frank recognition of human traits. North Americans were not brought up as conservers but as consumers. It is unlikely that the present generation will give more than lip service to the conservation ethic unless they are required to. An explicit way to effect this requirement is through people's pocketbooks.

Full-cost pricing aims to use the profit motive, operating through the market mechanism, to reward nonpollutive or resource-conserving producers and to punish those producers whose procedures are either pollution intensive or wasteful in social terms. Several steps must be taken to set these procedures into motion.

1. Government, in cooperation with industry, should compile input-output tables for the economy, regionally specific, in physical terms. In other words, we need an inventory—cross referenced—of what resources go into what production units and in what amounts. These then should be cross-referenced to already available dollar-flow input-output tables for specific years.

Besides recording inputs and outputs these tables should record residuals. They would then represent both materials and energy flow. In this way, an appropriate resource-distribution picture could be built up, based on national input-output materials-balance and energy-conversion tables.

Any residuals (solid wastes, heat, noise, etc.) showing in these tables would then need to be investigated by interdisciplinary teams to determine their effects on pollution and resource depletion.

In the above context, reduction in the stock of national wealth (induced by destructive consumption) might eventually and cumulatively lead to resource depletion. Again, it might not, especially if technological progress compensates for negative factors. In fact, future generations may be wealthier in this sense than we are.

2. To help clarify this question, we must have measures of the present value of future costs (and benefits) resulting from present acts. Precisely what these might be is unclear. It is likely that there would be several measures, depending upon which discount rates* were used in estimating present values of future costs and benefits. However, pollution costs of different natures accrue to the producer of a product, its consumer, and the general public. For each of these groups, therefore, different discount rates would apply. Of course, there will be many problems with using the interest rate for discount purposes. This

---

*A discount rate is the annual (compound) rate at which it is assumed a sum of money (the present value) will increase or decrease to meet an expected figure (cost or benefit) becoming due at a specified future date.

rarely reflects fully the true time (and/or risk aversion) preferences of society. Often it is a "functional response" to specific
government "monetary-policy" demands.

Equally, to know what the cumulative effects of combined
pollution on the environment in a given area might be (A and B
might not be harmful separately but together are deadly) will
likely be the single most difficult step in the whole job. This
requires an intimate knowledge of industry, demographics, and
other factors. But sensitive measuring devices, cooperation
from industry, and effective computer systems analysis offer
considerable hope.

Many controversies are bound to arise and there may be no
perfect solution. However, much work is already under way.
Canada, for example, has eighty-four pollution monitoring programs. Moreover, the Canadian approach is compatible with that
of the United States. Over time, international compatibility is
eventually possible.

Notwithstanding the difficulties, which we fully expect will
be immense, the cost effects due to residuals must be assessed in
accordance with the currently available technology for pollution
abatement. The cost of resource depletion would need to be
computed in accordance with present resource usage and its
reserve availability.

Once this is done it would be possible to have a fairly specific
preview of resource availabilities, the energy involved in their
transformation into goods and services, and the amounts of
postproduction and postconsumption pollution and/or waste resulting, or caused, thereby.

3. Finally, incentives should then be offered by government,
through the use of what might be called E (externality) taxes, to
those firms which are environmentally or resource-use frugal
either (1) as a result of substituting recycled (secondary) for
primary resources in production or (2) as a result of developing
and using technical capabilities which minimize the "potential"

occurrence of postconsumption residuals.

The whole purpose of the E tax is not to collect taxes from industry but *rather to stimulate industry to undertake those actions which would result in its avoidance*. The E tax would be an incentive tax so structured as to reward the nonpolluting firms and penalize the polluters. Indeed, it would become a recognized cost or credit of doing business.

This type of selective incentive taxation can certainly alter production costs. Increased charges (with resource-profligate or pollution-intensive production processes) would be passed on to the consumer in the form of higher-priced goods or services. Some people will choose to pay these higher prices. Others, who opt to use less pollution-intensive or resource-profligate goods and services, would pay lower prices because of what would, in effect, be a tax rebate for the producers. The benefits here would eventually be passed on to the consumer in the form of lower-priced goods.

It is well to remember that sellers are interested in their market share as well as their profit margins. Hence, a selective externality tax—an incentive tax—can redirect market forces. It need not, and would not, necessarily impede these. It would make it more profitable not to pollute and to develop resource-conservationist technologies and recycling industries, and even more profitable to be frugal with primary resources, including especially energy. An example illustrates how this might arise.

Mining and mineral processing are very energy intensive, especially in the thermal form, e.g., smelting. All this implies a propensity to pollute. To avoid this, hydro-metallurgical (i.e., washing) rather than pyro-metallurgical (i.e., burning) techniques should probably be used. Although there frequently is a high mineral content in slag residue from hydro-metallurgical processes, it is difficult to recover precisely because it is in slag, and also uneconomic given current cost/price structures of the industry. Hence, the incentive now is to use pyro-metallurgical processes. With E taxes added to pyro- and subtracted from hydro- processes, the latter would be in a more viable economic position.

Besides promoting the shift to processing methods with high recovery rates, E taxes could be expected to cause great interest in the secondary (recycled) materials business on the part of primary (or virgin) materials processors. This would be logical since these operators are already well established in the materials markets and thus their entry into upgrading residuals (or secondary materials) would represent only a cost—and not a marketing—challenge. Some new technology will be required to mass-produce secondary resources, but who is better qualified to generate this technology than present mineral processors—if given the incentive of the E tax?

Even a partial implementation of the E tax system and the resource inventory proposed here would go a long way toward bringing about a more satisfactory flow of resources than we have at present. This is because there are a few major industries in North America which contribute to the bulk of total pollution, and a few industries which are flagrantly wasteful in resource use.* These are the first ones that should be encouraged to move toward the use of secondary (recycled and/or recyclable) rather than primary-production inputs.

In conclusion, full-cost pricing would foster socially desirable growth by providing for a better understanding—in monetary terms—of each other's concerns by both industry and government. The shorter-term competitive survival of an industrial company would thus become compatible with the longer-term objectives and outlines of government planners. Moreover, much of the uncertainty (and the sheer bureaucratic waste deriving from this uncertainty) that faces the business world today

---

*One thinks in this context particularly of iron and steel making, automobile manufacture, nonferrous metals (aluminum), forest products, packaging, and petroleum chemicals.

about government goals and actions would be eliminated, as this greater coincidence of objectives is attained. Heaven knows, we surely need this. We also need the stimulus and discipline of a reinvigorated—and more truly competitive—market system to provide that creativity which will enable us all to get a great deal more out of living but in the process use up far fewer of our resources. Full-cost pricing offers precisely that opportunity.

13. THE OPTIMUM-MIX ECONOMY

The aspect of the conserver society that is most likely to become a focus for controversy is the jurisdictional question of who is to bring it about, the market or the state. The fears of more statism are indeed well grounded because public bureaucracies have an uncanny tendency to self-perpetuation and unlimited expansion—itself a source of great waste. There is also the fear that some Big Brother, some self-appointed council of sages, or some powerful bureaucrat or politician will govern our lives for us and threaten our individual liberties.

At the other extreme are the defenders of the market system come-what-may, right-or-wrong, good-or-bad. They bring to the defense of laissez-faire a religious fervor reminiscent of the great Adam Smith himself and seem to imply that the market has supernatural powers—nay, is God himself. There are enough documented histories of market failures to explode this myth and to suggest that the market is neither God nor the Devil but a human institution subject to the same vicissitudes and weaknesses as all human institutions.

When the market *vs.* state controversy becomes highly ideological, it is rooted in a priori assumptions, superstitions, and incendiary statements leading to a dialogue between the deaf and the mute. Besides, the protagonists and antagonists enter the argument with the firm resolution not to change their minds under any circumstances. To avoid that blind alley, we propose an investigation of the optimum-mix economy, based on pragmatism rather than ideology: the market, like the state and like conservation, is a means to an end, not an end itself. The end is human fulfillment, qualify of life, individual freedom.

Business Opportunities in a Conserver Society

How can individual business people prepare for and manage the challenges of a resource-scarce rather than resource-abundant future? In a society which responds to ecological imperatives, how can business turn perceived threats to its survival into opportunities? Indeed, need the conserver society be one of doom and gloom for business or might it not, in fact, turn out to be one of profit and boom?

Energy Opportunities

There is no doubt that in the next few years energy from conventional fossil fuels will become more expensive. In straight dollar terms, energy production costs will rise as long as less and less accessible sources—e.g., the tar sands—are used. A similar situation is likely to prevail for electric-power costs. Consequently, as prices rise, there will be great opportunities for businesses that specialize in helping to reduce these costs.

Industries which can provide energy-efficient technologies are bound to grow. Similarly, in-plant energy-conservation programs—and specialists in this field—can look forward to a bright future. In the energy context, a dollar saved can well become more than a dollar earned in the future, especially given tax burdens.

In the short term, then, before coal as an alternative energy source and certainly before any movement to solar, wind, geothermal, tidal, or any of the more distant energy sources is adopted, there will be a strong demand for energy-efficient technologies and also for a labor force skilled in their application.

At the plant level, new systems which prevent heating and cooling losses will be encouraged as they become economic. So too will retrofitting of machinery and equipment to eliminate unused or underutilized energy. Firms specializing in both activities will spring up. Engineers will be encouraged to apply their knowledge to design changes to make processes energy efficient. Consulting engineers will thus have immense opportunities to exercise their acquired skills here, and to develop new ones.

More insulation of industrial buildings—to avoid heat loss—will become practical. This will stimulate not only development of new insulation materials but whole new processes and applications for existing products. Secondary manufacturers specializing in the provision of such materials and their industrial and residential installation will be established.

As industrial firms begin to consider the life cycles of their energy inputs, as they undoubtedly will in an era of expensive energy, a variety of changes will occur in industrial building design; in warehousing and storage; in refrigeration; in inventory handling; and in transportation and delivery systems.

New standards will develop, with energy savings over the plant's expected life offset against higher initial capital construction costs.

Inventory-control systems will need to be redesigned; perhaps their size will be considerably reduced, especially if capital is tied up therein because of energy needs. This in turn could lead to the development of entirely different warehousing systems than we are used to. The frozen-food business, for example, might undergo radical change and this in turn might then substantially affect much of contemporary food processing and distribution. Alternatives to high-refrigeration-cost items will need to be developed. There is great scope here for business enterprise and potentially profits could be high.

Similarly, as we mentioned earlier, firms might look seriously into the prospect of reducing energy costs through renting rather than owning equipment, machinery, transportation, and so forth. If a thoroughly energy-efficient rental system is introduced it will have the distinct advantage of spreading the initial costs out over several users and also of maximizing use, thereby minimizing the down-time now built into all systems to handle peak-load situations.

There are many advantages to rental systems from the cost-effectiveness viewpoint. To return to the inventory situation, there could be cooperative, central, energy-efficient warehouses; scale economies could operate here to all firms' advantage. Firms could rent inventory space and refrigeration rooms. From these central depots delivery systems could fan out through the city and multiple deliveries could be made on one journey.

Granted, all this would take more organization than is now the practice and initially there might be foul-ups. But here lies a real challenge for managers. How do you make these new systems work? Again, specialists in these fields could command high salaries. Specialist firms could expect high profits. Perhaps they could work on a commission partially based on energy savings realized. There are a hundred and one variations of the theme, and the opportunities for business are indeed only as limited as is the insight of entrepreneurs.

Recycling and Resource-Conservationist Technologies

As we have stressed, resources are being continuously changed in both production and consumption. There are intended outputs (the car, house, artistic concert, whatever) and a variety of unintended by-products (the thermal pollution which accompanies a pulp mill's effluent discharge; the six pounds of garbage per day per person which Canadians discard).

There are, in other words, a variety of social costs and consequences to our current production and consumption practices which are not accounted for either in the profit-and-loss statements of firms, by government in their planning, or by individuals in their daily lives.

Pollution abatement after the fact is costly. If we encourage the recycling of postproduction and postconsumption wastes, we not only reduce the potential for pollution but also use our resources much more efficiently. If this is so, why isn't recycling already widely practiced, and why do those firms now in the business frequently not last too long?

There are several reasons.

Until a few years ago, virgin resources were thought to be abundant and were relatively readily accessible. Depletion and depreciation allowances encouraged the growth of extractive industries and, in simple dollar terms and largely because of the scale economies associated with this process, it was simply cheaper to use virgin resources in production than to use recycled materials.

It is becoming less cheap to do this now. But, in many instances, the full costs of using virgin resources are in no way yet properly accounted for. Today, the higher capital expenses associated with the extraction of less-accessible resources *are* being recognized. But the social costs of environmental degradation and postproduction pollution are not yet fully appreciated.

As far as postconsumer waste is concerned, the story is similar. Only recently have the full costs of garbage collection and its disposal through antiquated land-filling techniques, which of course take much of the best land around cities out of potential agricultural production, come to be recognized as real costs. To cover land with nonbiodegradable plastic and chemical by-products imposes a burden on us all; we are only gradually beginning to realize its extent. The story is similar too as far as our water supplies and other fundamental necessities of life are concerned.

Despite all this, we are slowly coming to the realization—as we have with energy—that there are more effective ways of using and disposing of resources. This is why recycling industries *now* hold out the prospect of substantial profit and very considerable growth.

Change will not come about, however, without a thorough reappraisal by governments of their positions as far as recycling industries are concerned. In this context, we have proposed that, in the future, the federal and state governments use their taxation capacities to give a real incentive to recycling industries. To give them, in other words, a significant tax break. Governments need to do essentially the same thing to encourage the growth of resource-conservationist technologies. There may be differences in detail here, but not essentially in approach.

Such an incentive tax approach would have three major, and relatively quick, effects in encouraging both recycling industries and the growth of resource-conservationist technology.

First, a boost would be given to present recycling industries. Their operations would become more competitive, and by the amount of the tax benefit. They could pass this saving on to consumers in the form of lower-priced goods, thereby increasing their market shares. They would then be better able to enjoy the scale economies which come through larger size.

Second, the tax incentive would encourage the movement of people and ideas into the recycling sector. With the prospect of substantial profit forthcoming, resource-conservationist technologies of all types would be encouraged rather than discouraged, as is now so often the practice.

Third, many of the contemporary primary-resource users—if their competitive disadvantages become great enough through such tax burdens—would themselves be encouraged to seek out alternative technologies and production processes. Not only would they switch to greater use of recycled inputs and thus further help these industries, but they also might get the recycling business themselves.

In social terms this would be advantageous. It is in our larger firms that we frequently find our most talented scientists and engineers, who are attracted by the high pay. Much of this talent is now underutilized, in a social sense at least.

Expensive scientific time is now too often spent on minor product differentiation rather than on systems innovation. If top management—through a tax system such as that proposed—begins to realize that there are significant cost disadvantages to staying with present technologies, they will begin to press for system redesign.

In this context the abilities of our engineers and physical and social scientists can be fully utilized. It will become profitable for management to unleash these talents toward the solving of some of our environmental and pollution problems. Thus rewarded for their efforts, and given technical and other support personnel and facilities, we could see immense and beneficial changes for society being brought about by these specialists, and in short order.

Right now the scales are stacked against the use of some of our best talent in this way. With an effective tax stimulus we would get not only appropriate technology but a rewards system for those inventing it, which would also be appropriate. Moreover, under any such industrial approach, we might at last begin to use the talents of our younger scientists and technologists many of whom are now unemployed or underemployed. This is a pathetic waste of human resources, especially considering the past investment in their education.

Finally, and where appropriate, solutions could be exported abroad. In this way some of any initial losses sustained by moving toward a recycling and less resource-intensive economy could be recaptured. At a different level, such appropriate technology could be offered to developing countries as part of any aid package, most of which would be much more suitable to their real needs than that which we now make available. And, as far as profits here go, we all are aware that foreign-aid budgets remain high. These amounts are unlikely to decrease over the next several years. The form of this aid could change radically, and to the greater benefit of the recipients.

Marketing and Conservation

We have been promoting the benefits of recycling and the development of resource-conservationist technologies basically in terms of cost, price, and profit. However, considerable public inertia still must be overcome, especially in the short term. This is why there will be a demand for the marketing of conservation and the principles of the conserver society, especially in the near term.

We associate disposability with affluence. How many pop bottles, Kleenexes, ballpoint pens, or disposable razors have you thrown away today, this week, this month? If you are like most North Americans, quite a lot. Moreover, the media are continuously insisting that we throw away more.

We can resist much advertising. Indeed, we often become so annoyed with the banality of TV commercials that we deliberately won't buy Brand X but instead would choose Brand Y—the one the commercial for Brand X panned. But, obviously, advertising (TV, radio, billboards, and all other forms) still sells.

Advertising is a multi-billion-dollar business. If it didn't perform for its sponsors, there would be no sponsors. So it is a very strong force—for good or ill—in society. Can't this persuasive and very effective medium be used to get the conservation message over? It can. Moreover, there are profitable opportunities aplenty for existing advertising and other promotional agencies to turn their talents toward selling conservation and conserver goods and services.

Even though the economics of energy and the depleting of natural resources might dictate a change from a consumer to a conserver society, it is necessary that a massive transformation of attitudes, beliefs, values, and behavior, certainly in the short term, occur before any changes *are* accomplished.

Every economic organization, other than the isolated self-sufficient family unit, needs an effective distribution, exchange, and communications mechanism. In complex societies like ours, marketing moves people, goods, services, and ideas from one place, or form, to another place or form. Marketing is a provisioning technology; its techniques and strategies can be used, and are used, to stimulate and support change. Precisely because of this, marketing in any complex society, consumer or conserver, will be fundamentally necessary.

What are some of the specific areas of opportunity?

1. *Marketing: Responsible Consumption.* As we have suggested, there will be immense opportunities for those in marketing to use their proven techniques to bring about reform in our collective inefficient consumption habits. Media and other programs to inform and persuade us that it is to our personal benefit to drive smaller cars, use public transit when we can, conserve home heating fuel, insulate our houses, make sure we buy goods or consume services that are energy efficient— will be in great demand. In this context a panoply of techniques can be used. Surveys will need to be undertaken. Results will need to be analyzed and effective promotional campaigns developed. The initial client for these services probably would be government. As they catch on, however, the private sector—in its own interest—will also become a significant source of revenue for such market research and advertising activity.

2. *Marketing: Recycling and Reverse Channels of Distribution.* Although many of the technological capabilities already

exist to recycle much more of our postproduction and, especially, postconsumer wastes than is now the practice, there is a variety of social obstacles and barriers to the acceptance of a recycling philosophy. The most significant of these is that at present there are no really effective reverse channels of distribution, no mechanisms to bring back from consumers what otherwise is thrown away. This is where marketing can play an immense role in the future.

One simple illustration. Right now the labor costs associated with separating and sorting garbage are frequently too high for firms to consider this option seriously. The amount of energy and materials constituent in wastes is also very high. Marketing can induce consumers to perform this function. It can encourage people to separate and prehandle their garbage, to take their cans, bottles, and paper to central depositories. A whole panoply of accepted techniques is available. Firms can devise cash bonus and benefit schemes, provide informational advertising and a variety of other approaches. Marketing management, in other words, can adapt its proven forward distribution techniques to create reverse distribution channels, all to the net benefit of society and, what is more, can make a profit out of it.

3. *Marketing and Packaging.* The packaging industry in North America has huge annual sales value, and it consumes enormous amounts of paper, glass, metals, wood, and plastics. To many it is seen as a source of great waste, but there are few who could argue that all packaging is unnecessary. It can reduce spoilage and certainly increase consumer "convenience." In an age of scarcity, the marketing manager will have his or her job cut out to develop, design, and implement new approaches to packaging and to physical distribution.

We will still consume in a conserver society. What we consume, and whether or not this is packaged and distributed in as ecologically benign a way as possible, will provide a challenge of unprecedented proportions.

4. *Marketing and Product Upgrading.* For the past thirty

years or so mass production to survive has required mass consumption. In the process, products have been built to wear out quickly or become obsolete, as fashions have changed or minor product or style innovations have occurred.

In any age of scarcity, a new challenge for marketing managers will be to convince consumers to purchase products—at higher prices—that last longer and are of better quality or have several uses. The consumer will have to be reeducated. He or she will have to be encouraged to pay for maintenance and repair services which increase product life. Similarly, consumers will have to be informed—through advertising and marketing—about system changes. New concepts, such as transportation networks rather than single cars, as *the* only viable means for individuals to get from place to place, will need to be explained.

Improved quality and increased versatility, through incremental changes to products instead of the introduction of completely new ones, will also need to be explained—marketed if you wish—to consumers

5. *Marketing: Nonprofit Institutions and Government Agencies*. In recent years, marketing techniques have been used to sell birth control, nutritional standards, flood and disaster avoidance, hospital programs, and a variety of social services. The need for this special kind of marketing is not likely to diminish in a conserver society. Indeed, in the early stages, it will probably increase.

Similarly, governments at all levels will require effective advertising campaigns if they are to sell a conserver society. Here lies the opportunity for great specialization, concomitant profits, and the *growth* of many specialized individual firms.

There will be lots of money to be made by the marketing people in the transition to a conserver society. In addition, there might be considerable nonmonetary rewards for them. The psychic satisfaction of selling something you can believe in is great. Individuals cannot and should not, of course, be expected to live on psychic income alone. But in a conserver society this would assume a greater significance to the individual than it does today.

The Responsibility of the State in a Conserver Society

The need for governments to play a role in the construction of a conserver society stems from potential limits to the capacity of "the market" to conserve.

1. *If the market does not normally account for externalities, the state must.* As we mentioned earlier, an externality is an unintended side effect either beneficial or noxious. Assume for instance that firms A, B, C, and D each produce a by-product, a, b, c, and d respectively, which is discharged into the atmosphere. Each of these effluents singly is totally harmless, but together, because of biochemical reactions, they are deadly. In such a circumstance no automatic market mechanism will correct that situation until it is too late. Only some form of state regulation will prevent disaster.

More generally, given the high interdependence of our activities and given the fact that our planet and certainly our continent are more and more crowded, some house rules must apply. A person living like a hermit in the country is free. A person who shares quarters with others must constrain his or her freedom lest it interfere with that of his or her roommates. From this realization of interdependence emerge house rules. On the collective level, the existence of externalities justifies the responsibility of government to impose house rules, which may include full-cost-pricing schemes, externality taxes, and the like.

2. *The state may have to intervene to preserve individual liberties.* This may seem paradoxical. Surely the state threatens individual rights. In fact, this is not always so. The government is an elected elite which quite often must be mobilized to protect

the individual against such powerful unelected elites as giant corporations and giant unions.

It is important to distinguish between the rights of the individual and the rights of the corporation. Sometimes the former fall victim to the latter, and in defending the market one may not always be speaking of the interests of the individual. Consumer sovereignty is gradually being eroded by producer sovereignty. The leviathans of enterprise may by their sheer size force decisions upon the consumer. How protected would the consumer be against rusty cars, substandard appliances, even fraud without some government involvement? The argument claiming that competition will protect the consumer against inferior products collapses when we realize that in a typical market situation with no state intervention (such as antitrust laws) competition breaks down. The reason is simple. The intelligent producer has every interest in the world in cooperating with other producers to fix prices and reduce product quality unless forbidden by law to do so. A host of oligopoly models in economic theory attest to that fact.

A perfectly competitive world may indeed be highly conservationist, but also it is as remote from reality as a perfect triangle—they are both logical constructs, not descriptions of experience. The real world is made up of oligopolies and giant production units. For the sake of both conservation and equity, the representative of the people, the legitimate government, must assume some responsibility.

3. *The price mechanism is unfortunately unreliable*. The classic free market theorem argues that the price mechanism alone can ensure conservation. The proposition runs as follows. If petroleum is in danger of depletion, let the price rise in order to discourage consumption. As long as the price is not high enough people will tend to waste the product, but if charges indeed do rise, a natural equilibrium will be reached. Therefore, advise the free marketers, laissez-faire the market for best results.

For the sake of clarity it is worth repeating the basic problems associated with the free market. There are four causes of unreliability of the price system as an economizing device.

a. *Price distortion due to market imperfection.* Most of the prices we pay today are "fictitious" as opposed to "real." A real price is that which would be achieved under conditions of free and unfettered perfect competition. This, alas, is not the situation in the real world.

First, there are price distortions introduced by monopoly and oligopoly power within markets. Oil cartels or multinational corporations may manipulate fees. As a result we never know for sure if the price increase reflects a real shortage or an artificial one.

b. *Price distortions due to subsidies and indexing.* Another distorting element stems from subsidized prices. With the complicated system of cross-subsidies characterizing our modern mixed economies, there are substantial supports that keep some prices artificially low—and conversely weighty taxes which keep other costs artificially high.

In addition we must note the distortion due to indexing of wages to the cost of living. As the nominal price of energy goes up, the cost of living index rises because of the important position of energy in all products. When the cost-of-living index climbs, unions ask for higher wages. The net result is inflation: most prices and wages go up but in a *real* sense the price of energy has *not* increased in proportion to wages and other income. The increase has been fictitious and therefore there is no conservationist effect. As prices go up at the service station, so also have wages, and the consumer uses the same amount of gas as before.

c. *Prices based on insufficient information.* We do not have, at present, full-cost pricing. Much information is lacking and cannot be reflected in prices. For example, the pricing mechanism does not distinguish between renewable and non-renewable resources, an eleventh-century Norman Church and an aluminum factory. As Schumacher has persuasively pointed out, the price mechanism is qualitatively blind because it translates everything into dollar terms even when such action is absurd. Witness, for example, the practice in our courts of law of compensating an accident victim with a sum of money for a lost leg or arm or eye. Obviously, the qualitative loss of an eye goes far beyond its dollar value.

d. *Prices operate with a time lag.* Finally, the price system is a communication network of signals and interpretations which may suffer from dangerous time lags. There are, after all, many steps in the pricing process: information gathered by the producer; information conveyed by the producer as a pricing signal; the signal received by the consumer; the consumer acting upon the signal.

At every step there may be delays that could lead to an overshoot-and-collapse situation. For instance, the price of an exhaustible resource can increase too late in the game to actually save that resource from exhaustion. As a result the system will have overshot its boundaries and crashed to disaster without the benefit of a gradual braking mechanism. The image is that of a car driven by a driver with slow reflexes. By the time he has seen the obstacle and decided to apply the brakes it might be too late to avoid the collision. The car stops eventually—after the crash—but the existence of the brakes as part of the car's equipment did not prevent the accident.

By the same token, the slowness in signal emission and response in our modern complex markets may well invalidate the "braking" effect that we would like to associate with a perfect pricing system.

4. *The size and risk involved in many industrial ventures invite government participation.* The great industrial projects that have developed on this continent over the last two centuries required some degree of state participation. The U.S. and Canadian governments subsidized railway building in the nineteenth

century through land grants, tax exemptions, low-interest loans, and interest guarantees. The profit expectations in the short and medium run were indeed very low and of a nature to dampen the investors' enthusiasm. It is an open question whether any railway would ever have been built by private enterprise alone.

The risks involved in today's industrial projects are substantially greater. For this reason the construction of pipelines and airports, the development of new airlines, hydroelectric projects, and so on, all require some government participation. This fact further demonstrates the need for an optimum-mix economy rather than either a purely private or purely statist one.

Some Principles for the Optimum-Mix Economy

We can now briefly describe how, according to the CS_1 assumptions, the optimum-mix conserver society would work.

First, the decision to allocate our activity to the private or public sectors will be made on the basis of the acid test of efficiency: getting the job done at minimum social cost. This implies clearly identifying the jobs to be done and full-cost pricing all visible and hidden costs. As a result, since activities now in the public sector may have to return to the private, and private activities may have to become public, professions of blind faith will have to be set aside in both situations.

Second, when it is judged that there is a useful role to be played by the public sector it must then be decided what level of government should intervene. In both the United States and Canada there are three levels: the federal, followed by state governments in the United States and provincial governments in Canada, and finally the municipality, which may be called upon to play a leading role in implementing many conserver policies. By the same token, we must not forget that some problems may have to be solved by cooperation between the U.S. and Canada and others on a truly international basis (OECD or UN).

Third, the optimum-mix scenario calls for prudent and subtle uses of public policy, of which there are at least three levels of intensity. At the first, which we can call moral suasion, the government suggests, recommends, enjoins but does not compel the private sector to do or not do something. For certain purposes, moral suasion with a responsible private sector may be enough.

At the second level, the government will use carrot-and-stick or incentive-disincentive measures to implement a program. The carrots may include tax credits, subsidies, or interest-free loans. The sticks may comprise surtaxes or penalties. The considerable inventory of fiscal and monetary policies is available here and may be used to significant effect.

At the final level is outright regulation. This may be backed by administrative decree or formal legislation and is the most powerful level of intervention.

It is one of the cardinal assumptions of the optimum-mix conserver society that the third level may be avoided if the first two are used judiciously. At all times, we must bear in mind that the state, the market, and indeed the conserver society itself are instruments to make possible attainment of the higher ends of happiness, satisfaction, or whatever. This ends-means distinction must never be forgotten.

III
CONSERVER
SOCIETY TWO

The Affluent Stable State

World View:
Moderation Is the Ultimate Virtue

Motto:
Do the same with less

Overview of CS_2

CS_1 was based on the idea of *efficiency* and did not challenge the ethic of accumulation.

CS_2 introduces the possibility of mild value change, namely, the acceptance of the notion of a ceiling to certain economic activities. Before we lose the confirmed growth maniacs, who find this notion repugnant, a simple idea should be stressed: growth is a relative concept. E. F. Schumacher, when asked his position in the no-growth issue, responded: "I have no position.... I mean it all depends. If my children grow, that is marvellous. If I start growing at my age, why that would be a disaster!"

The relativity of growth is also reflected in that of growth rates. To claim that a constant 5 percent growth rate in real GNP is ideal at any time is absurd. An analogy proposed by Tuzo Wilson at the 1977 conference of the Canadian Association for Future Studies makes that point very clear:

Suppose we suggest that there is some constant rate of growth that is ideal independently of circumstances. Let us examine such an "ideal" growth if applied to that of a human being. Assume it is 50 percent annually. If an infant at birth weighs 8 pounds at the end of the first year he will weigh 12, and at three years will only be 18 pounds. Obviously he is not growing fast enough! At four years he weighs 27 pounds, at five about 41 pounds, which is now quite reasonable! Continuing with the growth rate, he weighs 60 pounds at six, 90 pounds at seven, 135 pounds at eight and over 200 pounds at nine. The little fellow is developing into a monster....

The message is clear: the growth rate must be adapted to the size of the growing unit. In appropriate circumstances a very high growth rate is recommended, in others a lower one, in yet others a zero or even negative growth rate may be optimal.

As Wilson pointed out, the paradoxical fact is that, although everyone is likely to appreciate the good sense of this argument, a no-growth society is going to strike horror in the minds of some people. Why, he asks, do we recoil at the mere mention of a no-growth society while at the same time we do not point an accusing finger at a thirty-year-old adult and call him "a no-growth human"?

The attractiveness of a stable state is really dependent on two perceptions. First, we must determine at what *level* we are going to introduce zero growth and freeze the situation. Here it is obvious that the popularity of a low stable state will not be the same as that of a high. The definition of the level is most important and CS_2 explicitly favors a high level. Second, we must determine *where* zero growth will apply, across the boards or selectively. A selective freeze is of course the most reasonable choice.

CS_2, then, is "doing the same with less," with the promise that what we will call "the same" is negotiable, although the principle of a ceiling would have to be accepted.

In the chapters that follow, we shall first describe the rudiments of the stable-state philosophy and then outline a strategy involving many Z's.

ZANG (Zero Artificial Needs Growth)
ZIG (Zero Industrial Growth)
ZUG (Zero Urban Growth)
ZEG (Zero Energy Growth)
ZPG (Zero Population Growth)

A Message from Mr. Middleton

Mr. Middleton was unusual among the Sammy Squander crowd in that he was really of the previous generation. When they first met him he was the only mature student in the freshman year of business school. He had left an unpromising career as an associate professor in a fairly respectable Midwest college, having taught classics with mild success for about twenty years. Latin and Greek had appealed to him even as a teenager, and they were the subjects he had chosen to study under the Veterans Bill after he had given creditable service in the Marines during the Second World War.

Mr. Middleton, as he was always called in spite of being well liked, merged easily with his classmates on his second time around in college. This was at least partly because of his appearance. He was the sort of person who had looked the same age, neither old nor young, since he had been about twenty. Another factor was his unusual knowledge—both historical and mythological—and his ability to translate precisely every quotation and motto which arose in Latin or Greek.

He had offered his younger friends a kind of stability and, without being oppressive, had kept them from carrying their wilder exploits to extremes. When he had returned with a diploma to the Midwest to take up a middle-management position in a fairly large firm they had been somewhat sorry to see him go. It was with mild affection that Sammy remembered Mr. Middleton as he read his postcard.

Hyannis, July 28

Dear Sammy,

We're having quite a good time here on our two-week trip to Cape Cod. Mary and I came with the kids and are enjoying the usual good weather. It's a pity that you couldn't join us.

I'm sorry to hear that you're having a tough time but maybe it's the turning point before you settle down. I hate to preach from my middle-aged, middle-class, middle-of-the-road viewpoint but I still say you can have too much of a good thing. The way I see it, when you overdo anything you don't have enough money, time, or energy for something else. Keeping on an even keel is the answer, I find. Remember Archimedes' fulcrum. But enough of the sermonizing.

Take care and keep in touch.

Morrie Middleton

14. THE GREEK IDEAL

Is there any precedent for seriously considering the stable state? Have there been in the past visions of a stable society in which growth in production and population continued to a certain point and stopped, leveled off? Is it an entirely strange notion or has it been expounded, examined, and virtually ignored or forgotten long ago? Is there anything relevant, for example, in the classical tradition? The prominent feature therein, as it shows itself in contemporary Western use of resources, is a generalized ambition for Roman opulence. Lewis Mumford, indeed, goes so far as to suggest that the mass-consumption society began in the Renaissance with the unearthing of Roman splendor and subsequent attempts to recreate it. The other side of the classical coin, the Greek, has imprinted our philosophy, our mathematics, and our art, but has, so far, made almost no impression on our "getting and spending."

The Greek ideal is probably best expressed in, and perhaps even largely inferred from, the qualities of ancient Greek sculpture: simplicity, proportion, and harmony. These components of the ideal are assumed, with distant hindsight, to have characterized all the best activities of ancient Greece.

They have not, in any sense, fallen into disrepute; if anything, the less attainable they become, the more admirable they seem. Indeed, simplicity, proportion, and harmony appear so remote from our society as to be mere abstractions having nothing whatever to do with our present way of life, which is, after all, founded on complexity, disproportion, and disharmony. It would be, therefore, either impossible or highly artificial now to formulate a new utopia, an ideal society, based on these ideas. What we can do is examine what the Greeks themselves viewed as an ideal society, to see what aspects of it would seem desirable for our own future, and especially in this context, to examine its conservationist features.

Perhaps the most explicit expression of this approach to life as it would manifest itself in a utopian design is found in Plato's *Republic*. There we see laid out a set of conditions under which the sculptures' qualities can be translated into action—or at least those conditions which Plato could envision. Given our very different social values and much more advanced technology, some of the mechanics of the scheme are both unacceptable and unnecessary. What is more immediately relevant and interesting is the correspondence between the evils *The Republic* was designed to avoid and those which threaten our present society; and also the fact that Plato's utopia is a certain type of conserver society. Conservation was not a dominant theme in Platonic thought and scarcity was not a problem which Plato directly addressed, but a close look at the development of *The Republic* shows that these subjects were exactly what Socrates and his interlocutors were talking about. What we must ask ourselves is, to what extent are the classical virtues of simplicity, proportion, and harmony consonant with conservation? Is the apparent relationship between an aesthetically appealing way of life and the wise use of resources merely incidental or is it essential?

It is only by examining *The Republic* itself that we can approach an answer to these questions. Internal scarcity in the ideal commonwealth is initially avoided by locating it (at least implicitly) in the rich valley of Athens. Here, from tree-covered mountains through fertile valley to sea, springs every known natural resource. This bountiful land would provide the economic base for the healthy life of a prescribed number of citizens—something more than five thousand. Socrates (or

Plato) was of the opinion that men are bound together in a community by their mutual needs. An optimum population, consistent with unity of the state, is arrived at by computation from the number of occupations, and people in them, required to satisfy those needs. If all is going well, each person will be employed in the task for which he is most naturally fitted. So far, so good.

The problem of scarcity obtrudes itself only when unnecessary needs begin to proliferate. In describing the antithesis of his Republic, Plato envisions the disastrous consequences arising from men's desires extending beyond the real necessities of food, shelter, and clothing. He sees, first, that the community must be enlarged, "swollen up with a whole multitude of callings not ministering to any bare necessity: hunters... artists... actors, dancers, producers; and makers of all sorts of household gear, including everything for women's adornment... servants, barbers, cooks, and confectioners." He even foresees the need to increase the number of physicians to treat the inevitable sickness arising out of overconsumption. But these are problems of an intermediate stage. The final evil in the inflamed state is that the land ceases to be able to support the rising level of consumption; the commonwealth must extend its territory, and war with neighboring communities results.

It is interesting that Plato sees overcrowding not as resulting from a natural increase in the population and not even as being of the spatial type. (If anything, he is anxious that eugenically-planned birth should keep pace with death.) His concern is, first, with what could be called overcrowding of the economy, the proliferation of needless goods and services, and, second, with the infiltration of undesirables to supply them. Of course, just as a cancer grows at the direct expense of healthy tissue in a human body, luxuries are supplied only by depriving a society or part of it of its natural needs, or alternatively by plundering its neighbor's goods. The relationship between the wealth and welfare of Socrates's citizens is a simple one: if the "wealth" consists entirely of goods and services meeting natural needs, then welfare is assured; if it contains anything else, then welfare is in jeopardy.

Can Plato make a worthwhile contribution to a discussion of the desirability of a stable state? His community may in many respects be considered seriously deprived, his genetic engineering plans outlandish, and his view of nature naïve. But none of these aspects of his utopia is essential to his basic thesis: natural needs themselves are stable; their satisfaction is consistent with a stable state. It is the unrestrained runaway growth and fragmentation of artificial needs (and the goods to supply them) that, beyond a certain point, lead to societal disease, to disunity, and eventually to war.

The same essential idea was put forward many centuries later by a social philosopher who was also, and not just coincidentally, a classical scholar. Writing in the mid-nineteenth century, John Stuart Mill's vision of a stable state was a response to his own question "To what goal is society tending in its industrial progress?" He perceived that the growth path of population and capital in the older countries such as England could not continue indefinitely but must eventually encounter a natural limit. His idea was that a society should prepare to be content with a *certain measure of scarcity* rather than be compelled, suddenly, to accept a more severe one by being brought up short in its industrial progress. Essentially, he was at odds with the previous generation of political economists, including Adam Smith, who would have equated the stationary state with poverty, and "progress" with prosperity.

In envisaging his stationary state, Mill recommends that there should be maintained that level of production of material goods which, given even distribution, would supply the moderate wants of a strictly controlled population. His plan was to head off scarcity, so to speak, by meeting it halfway. Population

control, although it is essential for his plan, is not worked out in detail but, once an optimum number of persons is established, Mill seems content to rely on their pruduce and frugality to restrain overall consumption. Some moderate legislation favoring equality of fortunes, e.g., limiting of gifts or inheritance, would help to ensure a more even distribution of capacity to consume. Technological progress—the improvement and/or enlargement of machinery—he thought should be used to "abridge labor" and not, simplistically, to increase production.

The idea of liberating the individual from laborious or distasteful occupation is part of Mill's conception of what human life should be, and springs from the same source as the "values" aspect of his conservation plan. Mill sought to halt the headlong rush of increasing production not only because of his rational recognition of ultimate resource limitation but also because he viewed the constant struggle for wealth as a less-than-human activity.

In the stationary state where "no-one is poor, no-one desires to be richer," a person would have room for "solitariness" and time to improve himself, to develop and enjoy his and his fellow citizens' artistic, scientific, and philosophical abilities. Indeed, this conception of the ideal state is one in which only those factors limited by nature need be conserved. Mill's estimate of humanity is obviously a much more sanguine one than Plato's, and his utopia therefore is seen as relatively easily achieved. Whereas Plato thought it necessary to control strictly all aspects of life, Mill seems content with a little supportive legislation, and relies on his fellow citizens to be as cooperative and reasonable as he himself.

It is especially interesting to examine the stationary state because it provides an unusual example of consumption control for what might be called "anthropocentric" ends. John Stuart Mill did not see the human species as just a part of nature, as bound to protect the environment, or any such thing. Nor did he view human welfare as proportional to the "grossness" of the national product. He thus stands outside the category of the efficiency-model builders of his own time. His unusual and largely disregarded proposal was to cease the never-ending growth of population and capital once a certain plateau of affluence had been reached.

A contemporary exponent of the high stable state option is the British economist Ezra Mishan. He has underlined the costs of economic growth and has advocated slowing it down. The level at which it should be retarded or completely stopped is unclear but, on one occasion, he described the optimum desirable level of industrialization as being that achieved in Western Europe around 1900. The first years of the twentieth century have indeed been viewed by many authors as a golden age. This was the period of the Belle Époque in Paris, the period of advanced scientific achievement, the full bloom of the Newtonian world view, the dawn of Einstein's relativity. The automobile was making its appearance, electricity was harnessed, and aviation was moving from the status of a sport to that of an industry. Transcontinental railways were criss-crossing the globe and intercontinental steamers were uniting the world. Upper-middle-class ladies were strolling the Champs Élysées and Toulouse-Lautrec, Paul Gauguin, and a whole generation of artists were marking the dawn of the twentieth century. It was the best of times because the human species, at least in the West, had achieved freedom from want through technology *without irreparably harming the environment*. There was optimum industrialization in the view of many. It was the worst of times only because the fruits of this industrialization were very unevenly distributed. If a classless society had existed in 1900, would it not have been the very stable state that reflects the golden mean? Mishan's insight is certainly food for thought and reflection.

Without going back to 1900, when few of us were alive, it is possible to make a similar argument with respect to the more recent past. If Americans were asked point blank whether they would agree to reduce their energy consumption by one-half, many would probably recoil in apprehension and reject the idea. Yet energy consumption in 1960 was about half what it is now. Most of us remember 1960. Surely we had a civilized country then, with roads, electricity, entertainment, and so on. Yet we were consuming only half the energy we are using now. Have we, by doubling our energy consumption, doubled our happiness? Still better, do we expect to double our present happiness by consuming twice as much energy in 1985? Surely there are grounds for skepticism and, thus, for an argument for CS_2—a high plateau of material affluence but not a steeply ascending cliff leading to eternal increase in throughput.

15. ZANG

or Zero Artificial Needs Growth

As we pointed out earlier, one of the principal sources of growth of our consumer society is the proliferation and fragmentation of needs. "Proliferation" means the constant expansion of existing needs and the creation of new ones. "Fragmentation" is the process of dividing and separating what were hitherto unified demands or desires. What was at one time a simple desire for cleanliness and hygiene has fragmented and proliferated to become a complex set of needs for every part of the body—from anti-dandruff shampoo to aerosol foot deodorant.*

Both fragmentation and proliferation are nurtured, amplified, and channeled by the visible and hidden persuaders of our society—the wealth of marketing, advertising, and quasi-advertising-type instruments now available for use by business and government. The result is need explosion, with the satisfaction of any one need immediately creating another in an exponential fashion. Since there is, in terms of satisfaction, a natural tendency toward diminishing returns, we require ever greater and more complex commodity packages to maintain a level of satisfaction. Consumer-oriented advertising creates permanent frustration by constantly attempting to widen the gap between what we own and what we wish to own. Our yearning power far outstrips our earning power, increasing the possibility of either great psychological alienation or financial collapse or both. It is

*For an elaboration, see William Leiss, *The Problem of Human Needs*, Science Council of Canada, 1975.

clear that the attainment and maintenance of a stable industrial state will require a greater control over the generation of new needs.

The classification and hierarchization of needs is a process that has been occupying sociologists and psychologists for years. Let us return to an earlier statement about the interface

Criteria for a Classification of Needs

Criterion	Categories
1. By Origin	Innate needs versus acquired needs (hunger versus need to smoke)
2. By Method of Satisfaction	• Needs that can be satisfied by primary economic activity (e.g., hunger) • Needs that can be satisfied by secondary economic activity (e.g., mobility over long distances requires vehicles such as cars) • Needs that require services (e.g., anxiety takes us to the psychiatrist)
3. By Interpersonal Impact	• Parasitic needs: satisfaction for A is at the expense of B (e.g., prestige) • Symbiotic needs: satisfaction to A requires satisfaction to B (e.g., speaking same language) • Neutral needs: no interpersonal impact
4. By Arbitrary Value Judgment	It is decided that some needs are "good" or "legitimate" and that others are "bad," "not legitimate," or "false"
5. By Intensity	Needs are classified by the degree of pain or discomfort that they cause independently of the source of need (e.g., need for heroin by an addict more intense than his natural need for food)

between psychology and economics by relating "needs" to their satisfiers' commodities. The object is to propose a classification of needs and commodities which will allow us to hierarchize them. We will then be able to set priorities on those that have to be satisfied before the others.

It is possible to classify needs by using one or more of the following criteria: origin, method of satisfaction, arbitrary value judgment, interpersonal impact, or level of intensity.

To classify needs by origin involves making the distinction between innate or so-called "natural" wants and culture-generated or artificial requirements, for example between hunger and smoking.

Needs can also be classified by method of satisfaction, that is, by the commodities involved in the attempt to relieve them. Thus, we can speak of agricultural needs (food), industrial needs (cars), or service needs (psychiatry).

The classification of needs according to interpersonal impact divides them into three categories, parasitic, symbiotic, and neutral. Parasitic needs, such as prestige, can be satisfied for A only at the expense of B. Symbiotic wants, such as speaking the same language, can be satisfied for A only if accorded to B at the same time. Neutral requirements have no interpersonal impact.

The last criterion for need classification is level of intensity. This means that, independently of their origin, certain needs would be judged to be more intense than others in the subjective estimation of those who experience them. In such a classification, cigarettes for the smoker and heroin for the addict would be placed at the same high level of intensity—and perhaps even higher—than the need for longevity. Similarly, the need for vitamin K, a natural need indeed, would be of low intensity insofar as it was not translated into immediate discomfort. Before we attempt a hierarchy of needs, let us look at the other side of the picture: the commodities whose role is to satisfy needs.

The table shown below proposes a classification of various types of commodities. As in the case of needs, commodities may be classified by method of production, by interpersonal impact and, unlike needs, also by their "conserver" or lack of "conserver" qualities. The point of this double taxonomy of needs and commodities is not intellectual gardening but rather the preparation of a justification for a freeze in the growth of certain needs, which is central to the strategy of CS_2.

In a CS_2-type stable state all existing needs and commodities at time $T=0$ will be considered legitimate and fully acceptable. These will include all the innate needs (such as the need for food, drink, shelter, sexual satisfaction, and so on) and the artificial or acquired needs (such as desires for pizza, cigarettes, chewing gum, yoyos, or red garters). Excluded from the list will be hazardous commodities or needs, whose satisfaction is antisocial or sadistic.

Alternate Criteria for the Classification of Commodities (Need Satisfiers)

Criterion	Categories
1. By Method of Production	*Primary:* Commodities from the primary or extractive sector *Secondary:* Commodities from the transformation or manufacturing sector *Tertiary:* Commodities or services *Quaternary:* With no visible seller
2. By Interpersonal Impact	*Elitist* commodities that can be available only for an elite *Public* goods that can be available for all by definition *Neutral* commodities that have no interpersonal impact
3. By Degree of "Conservation" or Absence of Conservation associated with their production	*Conserver* commodities are those that satisfy the requirements of the conserver society *Nonconserver* commodities are wasteful commodities

An attempt should be made to prevent the creation of new artificial needs unless they satisfy the "ZANG rule," which distinguishes between harmless new needs and harmful ones. This decision rule can be formulated as: No new artificial wants will be gratified unless (a) they can be satisfied by "conserver" commodities *and* (b) they can be satisfied for society as a whole and not exclusively for a privileged minority. A conserver commodity is either one that requires low throughput (is not energy thirsty, resource hungry, or highly polluting) or one that in spite of requiring high throughput is easily recyclable and utilizes primarily renewable inputs.

If a new artificial need can be satisfied by a conserver commodity, then there is little reason for society to prevent its birth at least on ecological grounds. But it may be threatening to society because of its elitist nature, hence the second part of the decision rule.

An elitist commodity is one which by its very nature cannot be made available to all. A yacht, the largest car on the block, prestige, power are all commodities which increase inequality. Their interpersonal impact is high and negative. One person's prestige and power are at the expense of everyone else since not everyone can be prestigious and powerful at the same time.

Insofar as some form of egalitarianism is desirable in the conserver society, viewed not just as a series of conservationist measures but also as a complete model of society, the two aspects of the rule must apply.

An example will illustrate ZANG and show it to be less radical than would appear at first blush. Our hero, Sammy Squander, is a child of the hi-fi generation and he has gone through the various stages of escalation of the industry. About twenty-five years ago, Sammy was sensitized, together with the rest of Western society, to "high-fidelity" sound reproduced through one speaker with a minimum of distributions.

Twenty years ago, Sammy became convinced that true enjoyment of music comes from the use of not one but two speakers. After all, the argument made by the advertisers was quite plausible: we have not one but two ears. If we have a speaker for every ear, then it would be possible to produce depth in sound with richness, diversities, and a feeling of "aliveness" not obtained through one speaker.

Five years ago, Sammy was gently persuaded that even stereo is not enough; only quadraphonic sound will do. "To fully simulate the experience of a live concert performance," the commercial declared, "it is not only necessary to achieve proper stereo separation but also to reproduce the background noises one hears at a concert. In a live performance the music bounces off walls, chairs, galleries and is transformed into background sounds. Only four strategically located speakers with the proper tweeters and woofers for faithful high-fidelity reproduction can achieve that effect." Armed with this new "knowledge," Sammy proceeded to buy the quadraphonic system.

Alas, now a year later, the clever advertisers are at it again. The subject is . . . octophonic sound requiring eight strategically placed speakers with tweeters, woofers, and the powerful amplifier to go with them. The cunning advertising ploy: four speakers may well cover the lower surface of the room, but what about the ceiling? For truly enveloping sound, one needs four speakers for the floor and another four for the ceiling.

At every escalation energy consumption goes up, throughput is increased, and our unquenchable thirst for nonrenewable resources becomes ever more pressing.

ZANG draws the line symbolically at quadraphonic sound. For those of us who have been trapped into developing the high consumer needs of today, CS_2 still allows for their satisfaction. But further growth is arrested.

Does ZANG involve a high psychological cost? The answer appears to be no. If you have not yet experienced the need for octophonic sound you will suffer little by not having it. However, for the sake of completeness, it is important to point out the possible lack of stimulation that may result if we do not develop new desires. In the literature on needs, reference is sometimes made to a basic, high-intensity, natural psychological factor called "need stimulation." Failure to satisfy our nervous system's apparent need for external stimuli may result in pathological sensory deprivation. This has been noted by Hans Selye, who identifies an intense need in the higher animals for *stress*. He distinguishes "eu-stress" or "good" stress, which we seek (the equivalent of stimulation), and "dis-stress" or "bad" stress which we wish to avoid.

For good or for ill, consumer goods act as stimulants for the nervous system, somewhat as toys do for the child. Depriving a child of toys may undermine the development of his intelligence. Similarly, an adult deprived of the stimulation of consumerism may well be "bored," "depressed," "alienated," and develop the very "distress" which Professor Selye counsels us to avoid.

Many middle-class people cheer themselves up by buying knicknacks, clothes, appliances, even new cars. The depreciation of the current car becomes symbolically linked, in the person's mind, to the aging process. Changing to a new car represents rejuvenation. To remove the props, costumes, makeup, clothes, and other paraphernalia which help create "image" may be a psychological problem and yet an ecological necessity. Therein lies the dilemma.

What, then, is the answer? The extreme solutions are either to cater to our psychological need for "toys" and accept the evils of explosive throughput, or to stifle and suppress that desire at the cost of psychological deprivation. If in fact it is true that unlike a horse, which just needs, a human being needs to need, or in other words his or her nervous system is a complex apparatus that is overendowed for mere survival and requires constant stimulation, then a freeze in new pseudo-wants must be qualified with something else: a phase-in/phase-out process.

A realistic balance of needs with commodities must be maintained. When we buy the "largest car on the block" it may well be for a deep unconscious motivation that has little to do with travel or vehicles. The "largest car" is a symbolic structure satisfying perhaps a need for prestige which itself expresses a deeper feeling of insecurity. The Big Rock Candy Mountain had the purpose of satisfying that apparent need via heavy throughput: produce the cars, the houses, and the various commodities that are required to confer prestige. The conserver way, at the CS_2 level, is to phase in direct satisfiers of the basic need for security, which, in the final analysis, may be much less costly in ecological terms than the alternative. Someone conscious of his or her intrinsic worth may not need the trappings of the big car, the swimming pool, and the other symbols of success. Herein lies the challenge of CS_2: to provide satisfaction by going to the direct needs and to freeze the growth of new artificial ones.

The Role of Advertising

ZANG is an objective of CS_2, not a strategy. Since the development of new artificial needs is aided and abetted by advertising, a freeze in their growth must also directly or indirectly affect advertising. Two attitudes are possible. The first is radical and calls for the abolition of all persuasive advertising; the second is moderate, instead calling for more even-handed advertising.

Abolition of persuasive advertising would imply a fundamental reorientation of our economic system, leading to a natural and probably very determined resistance from all industries associated with advertising. It would require a revolutionary change in attitudes and substantial social upheaval. On these grounds alone it is not recommended at the CS_2 level.

But in addition to avoiding unnecessary upheaval, there is another reason that would militate in favor of retaining advertising. The differentiation between "persuasive" and "purely informative" advertising is one of those hairline distinctions that cause great debate between linguists, psychologists, and communication experts. It is our view that, in fact, the distinction cannot usefully be made. Every communication of information conveys by its very medium a certain amount of persuasiveness, either positive or negative. As Marshall McLuhan has demonstrated, if the medium is the message, there is no neutral medium because there is no neutral message.

Rather than try to impose a distinction in such a hazy area, we instead propose the adoption of the second option. The conserver society should not be a new form of dictatorship but instead should fully espouse the principle of free choice. Therefore, the idea should be to optimize the information available to the consumer and to let him or her, not some self-appointed council of sages, make the relevant decisions.

There is reason to believe that, in the context of 1978, the information available to the consumer is not optimal because it is one-sided. The advertising industry is overwhelmingly committed to the selling of the Big Rock Candy Mountain.

The creation of needs is more often than not the result of an advertising campaign using familiar elements of motivation theory. These are principally scare tactics, sexual association, and image building. Scare tactics are usually in the form of "Your best friend is telling you you may have bad breath. Unless you buy mouthwash X, the friendship will be ruined." Sexual association resembles the carrot rather than the stick. It is sometimes subtle but more often overt. The third tactic, image building, consists of the familiar technique of focusing on the user rather than the commodity. For example, a certain brand of cigarettes is associated with "independent" or "strong-willed" people, or an after-shave lotion with the Japanese martial arts.

To counter this one-way advertising, we need a persuasive campaign in the opposite direction. In effect, the advertising industry, which has developed an important inventory of communication techniques, should be mobilized to outline the merits not only of the cornucopian but also of the conserver society. The practical way of achieving this is through even-handed publicity: *Follow every commercial that is developing a new artificial need with an anti-commercial attacking it.*

This is not now done. Although there are commercials that challenge existing products, the advertising is by no means even-handed. The commercial for the product is found in prime-time television (say, the Stanley Cup Finals or the Super-Bowl), whereas the anti-commercial is found in an obscure consumer magazine often not even available on the newsstands or on some equally obscure educational television station. Obviously such imbalance favors the proliferation and fragmentation of needs, not the slowing down of their growth.

When agencies promote new super deodorants, deep emulsifiers, rejuvenation creams, or "double-power electric back scratchers," counter advertisers should be there to blow the whistle. Or, with the introduction of octophonic sound, the counter-advertisements could point out that quadraphonic, stereo, or even one good speaker can do an excellent job at a fraction of the price. If the consumer still wants the octophonic it will be his prerogative to get it. His freedom will be increased, not decreased, by making both sides of the story available to him.

The idea behind even-handed advertising is similar to the one underlying two fully accepted institutions of Western society: the judicial system and political debates.

The judicial system is based on the even-handed advocacy of opposing causes. It would be unthinkable to claim that a trial is fair if only one lawyer were heard. Similarly, candidates for an election demand equal time on TV because they are fully aware of the potency of that medium. Why then should we have a one-way system in a field that is becoming at least as important for society as a fair trial and a fair election, namely the unlimited growth of needs, commodities, and their attendant by-products, pollution, environmental deterioration, and social alienation springing from increased inequality?

This of course is a difficult question. Two possibilities exist. Either provide a collective budget funded by the taxpayer to pay for the anti-commercial—a distasteful and inefficient method— or ensure that the original sponsor will in fact pay for both commercials. This may seem far-fetched but is not really so. If a sponsor spends $50,000 for two minutes of television time, then the first minute could fund the pro-commercial and the second minute the anti-commercial. The TV network would, by law, have to hire another advertiser for the anti-commercial. The effect would be to discourage announcements based on flimsy grounds but should not deter those that are sound. Moreover, even-handed advertising is likely to bring excitement back to TV: a constant Perry Mason-type show at every commercial, instead of the doldrums of dull publicity.

Before the reader gets left too far behind in what could appear as a much too bold initiative, it would be worthwhile to consider two points. First, even-handed advertising is proposed here only for those new commodities that are being introduced through publicity and that have failed the ZANG rule on either ecological or elitist grounds or both. It is not meant to be extended necessarily to all commercials (although a good case can be made for that possibility, along the lines of the principle of truth in advertising). Second, to attack even-handed advertising is implicitly to make a plea for one-sided promotions (unless one advocates the abolition of advertising altogether). Since no fair person would, on reflection, recommend showing only one side of the picture, we can conclude that ZANG and even-handed advertising are a plausible moderate venture perhaps going beyond our present system but certainly not tearing at the fabric of our society.

16. THE OTHER Z'S

Although ZANG is the principal objective of CS_2 there are other Z's that merit attention. In each instance the idea is not to roll back or change what is but rather to freeze further growth at a certain predetermined level.

Zero Industrial Growth (ZIG)

The Industrial Revolution brought with it untold benefits which have helped the starving masses of the world. At the CS_2 level, it is not a question of challenging what has been achieved but knowing when enough is enough. Applied selectively, ZIG calls for a freeze on industrial growth in certain sectors and continued growth up to a certain point in others.

This objective is likely to raise eyebrows, since it goes directly counter to the conventional wisdom of the past two decades, which has stressed the theme of "salvation through industrialization." The "value added" of industrialization is coveted by most countries and regions that wish to increase the processing activities within their geographical borders. Canada, for instance, has consistently complained that she is perceived by the international community as a raw-materials producer rather than an industrial nation. Canadian development plans (like those of many of the provinces within Canada and most nations of the Third World) stress the need for industrialization. It could appear strange then to argue for the very opposite, zero industrial growth. Yet, if a high stable state is indeed desired (and once again its desirability is something for the public to decide), zero industrial growth is essential *for those areas which have already achieved an industrialization optimum.*

Applying the principles of selective conservation, it would be necessary, in effect, to select regions, sectors, and activities for continued industrial expansion and at the same time to identify other regions, sectors, and activities for zero industrial growth. Between these extremes there could also be intermediate zones requiring only slow industrial growth.

Central to the notion of zero industrial growth is, of course, the concept of "optimum" as opposed to "maximum" industrialization. It would be foolhardy to assume that such an optimum could be easily defined.

The British economist Ezra Mishan, as was noted previously, has argued for a return to 1910, the period immediately preceding the First World War, when the Western world had reached a plateau of affluence but was not encumbered by excessive pollution, threats of resource depletion, or grave ecological imbalances. Although there were alienation and social strife, they stemmed more from inequality in income distribution than from the imbalance of man's relationship to nature. It was an apex of sorts, a viable society which did not breeed within it the roots of its economic self-destruction. That world did of course self-destruct, but it was political and military struggle that did it, not industrial growth itself.

The choice of a model is arguable but the principle remains interesting: there could be an optimum industrial level beyond which further growth would be counterproductive.

Zero Urban Growth (ZUG)

The argument for an optimum level of industrialization has an obvious extension in the field of urbanization. There is an optimum size for cities. Between the various levels of supercities of the future defined by the Greek futurist Doxiadis, namely Megalopolis, Ecumenopolis, Anthropolis, there is also *Entopia* or the "first" place (as opposed to utopia or the "nowhere" place). Entopia may mean different things to different people, but here again the CS₂ philosophy stresses that optimum is neither maximum nor minimum.

Entopia could take the form of a conserver city, a city that is the living embodiment of the philosophy of the conserver society.

There are three broad approaches to limit urban growth that may be considered.

1. Policies to reduce the growth of large cities by regulating and limiting building permits for commercial, industrial, and institutional activities within the city core; by prohibiting location, within cities, of highly polluting industries; by imposing a "residence tax" on offices located within a city in order to reduce the rural-urban migration; and by prohibiting private cars within the city center.

2. Policies to encourage growth of medium-sized cities, including subsidies for the infrastructure of medium-sized cities to make them more attractive; decentralization of provincial and municipal government activities; decentralization of university locations; development of a network of new towns around the high urban concentrations to relieve the city of many burdens; and development of towns in the North (Arctic and Sub-Arctic) to relieve the high density of population in the region of the U.S.-Canada border. (This development, of course, must not be allowed to interfere negatively with the native populations of the North.)

3. General policies to improve spatial organization; this will include improvement of intercity transport networks; improvement of intracity public transportation; and further generalizations of rental schemes in housing (con-

dominium, or other forms of communal sharing), transportation, business, etc.

Zero Energy Growth (ZEG)

There is a conspiracy among certain special-interest groups to persuade the general public that whereas hitherto there were two inevitable things in this world, death and taxes, now there are three: death, taxes, and growth in energy consumption. As was indicated in the discussion of CS_1, there is no overwhelming necessity for growth in energy consumption. The existence of energy resources becomes, in a twisted version of Parkinson's Law, our excuse to use them up. Supply creates its own demand. Because we have fossil fuels to use up (or we think we should use them up as soon as possible), we invent electric windows and six-way power seats in cars rather than first feeling an intense need for those gadgets and then expressing that need by a "demand" for energy. This requirement is often not spontaneous, but induced by the supposed existence of plentiful supply.

At the CS_1 level, energy conservation was achieved largely through technological fixes, RICH, and innovative resource-management techniques. No particular energy-conservation objective was identified in terms of an optimum energy-consumption level per capita.

At the CS_2 level there would be a peremptory objective of zero energy growth per capita, and a variety of policies would be introduced to respect that goal—even if it involves sacrifices—unlike CS_1.

Zero Garbage Growth (ZGG)

This is another objective that can be achieved once it is realized that there is no universal law of nature decreeing that garbage production per person must increase by 5 percent annually. Again at the CS_1 level, no explicit freeze on garbage creation is implied but only greater efficiency to eliminate unnecessary production. At the CS_2 level, we posit a definite objective of zero growth in this sector—and in the final analysis the sacrifices that will have been made may turn out to be very small.

Zero Population Growth (ZPG)

The objective of zero population growth is a conditional one. Unless there are excellent reasons showing that population should expand or contract, we may assume that the optimum is what we now have. As was discussed in Chapter 4 the whole issue must be viewed in terms of two coefficients: a "multiplicand," or the actual number of people involved, and a "multiplier," which relates to their per-capita consumption of various resources. If we arrest the multiplicand but not the multiplier, we may not have achieved much. If in the year 2000 the same population as in 1977 consumes twice as much energy per capita, produces twice as much garbage, and pollutes twice as much, these people are obviously twice the burden to the environment that they were in 1977. Alternatively, if between 1977 and 1995 population doubles but the multiplier decreases by half through the application of conserver society policies, we will not be worse off than in 1977.

Population must be examined together with life-styles, not as an idea in isolation. There are trade-offs between more people with conserver life-styles and fewer people with squanderous ways. Limiting population alone may not achieve conservation at all.

At the CS_2 level, ZPG is desirable together with the other conservationist policies. An exception, however, must be made for immigration. Immigration is one way of sharing this continent's resources with the rest of the world. To close the gates on the pretext of ZPG is to adopt a Lifeboat North America concept, namely that this continent is a refuge in a troubled world. The more people rescued, the greater the danger to the lifeboat's original occupants.

Although there is some truth in the assumptions of the lifeboat metaphor, to keep out drowning people may be not only selfish but also dangerous. The drowning people may well sink the lifeboat anyway out of sheer frustration. For this reason, a sound immigration policy for the United States and especially for Canada with its wider spaces must coincide with ZPG internally. The level of immigration could be governed by the absorptive capacity of the land.

Zero natural growth in native populations together with a large influx of immigrants is bound to lead to social problems. However, this continent has been built by immigrants and, with the exception of the native Indian and Inuit, we are all immigrants—with different arrival times. To challenge the right of others to do what we did, in spite of the problems, is morally untenable. A way must be found, at the same time, to limit the growth of population and allow less-fortunate people to share our resources.

A Maximum Per-Capita Income?

A maximum per-capita income is another CS_2 idea that is certain to receive mixed reactions. On the one hand, it is worthwhile noting that the idea is not so radical as it may at first appear. Sweden is thinking about it, Britain is not far from it, and even in the United States and Canada the workings of the progressive income tax are designed to minimize income differentials. If we set aside the many tax loopholes that are available to those who can afford to discover them, there is, in effect, a taxation system that severely discourages high income beyond a certain threshold.

On the other hand, limiting per-capita income is psychologically threatening to many people who see it as a disincentive to effort, creativity, innovation, or even work. Besides, it is argued, once the ceiling is reached on monetary income, all sorts of devices are likely to be introduced to reward certain people in nonmonetary ways. The Russians use the system of state-owned villas on the Black Sea, luxury apartments, coupons for special department stores, ballet tickets to reward the "more equal" members in their "egalitarian" society (usually hockey or chess players, athletes, politicians, etc.). In China, noneconomic but quite tangible prestige commodities are bestowed on the elite, although the monetary income is kept at a level compatible with the average.

Rather than an absolute ceiling on income, a more palatable alternative might be a moving scale of acceptable income differences between the lowest- and the highest-paid members of society. This may mean a $5,000 to $50,000 range (in 1977 prices) or even, say, a $5,000 to $500,000 range if those income differentials are deemed to be acceptable. In general, however, the CS_2 philosophy would urge us to assure the public of a basic *minimum* income, which could presumably be given in terms of a negative income tax, coupled with a *moving ceiling*, which would be defined as a multiple of the minimum income. If the multiple is 10, then that will be the acceptable range of differences. As the floor income goes up so can the ceiling income.

To sum up, the stable-state ethic of CS_2 must be given a wide rather than a restrictive interpretation. It should be viewed more as a rule of thumb, whether we talk about ZANG or the other Z's, not as a divine commandment.

Based on the reasonable principle that there are *optima* which are not always either *maxima* or *minima*, the steady state calls for the charting of a middle course between extremes—and moderation in the very application of this laudable principle.

IV
CONSERVER
SOCIETY THREE

The Buddhist Scenario

World View:
Being, Not Buying

Motto:
Doing Less with Less and Doing Something Else

Overview of CS_3

The Buddhist sees the essence
of civilisation not in a multiplication of wants but in the
purification of human character.
E. F. Schumacher

With this phrase the late E. F. Schumacher epitomizes the nature of our third conserver society: the Buddhist Scenario. Our
aim here is not more efficient growth (CS_1), or achievement of
an affluent stable state (CS_2). Both our previous scenarios leave
unquestioned most of the principal tenets of our industrial society and still function within a paradigm of material accumulation. CS_1 and CS_2 do not demand radical changes in our values
or in our daily lives. CS_3 does.

We gave CS_1 the slogan "Doing More with Less." While
advocating a reduction in the physical throughput of material
resources and energy and suggesting that we recycle many of
the unintended by-products issuing forth from production and
consumption, our orientation was toward improving the efficiency of a growth-oriented society and economic order. In
depicting our second conserver society, we introduced the idea
of a potential ceiling on material affluence. We did not specify
where this ceiling should occur, but left this to the wisdom of
society. We were concerned with how to distribute the fruits of
affluence in a just and equal manner once we had decided upon
a ceiling level rather than with whether we should continue to
search for a higher standard of material life.

To continue our metaphor of the Big Rock Candy Mountain,

in CS_1 there's always another mountain to climb, while in CS_2 we are thinking that, after the next mountain in the range, we might stop after all and let the rest of the climbers catch up with us. In CS_3, we ask ourselves whether we should be climbing mountains at all and we answer: "No, it's gotten us into a real mess already. Let's do something else that we find more satisfying." Hence, the slogan of CS_3: "Doing Less with Less and Doing Something Else." The "Less" refers to less material consumption while the "Something Else" means developing the spiritual, emotional, creative, and intellectual potential which we all have as human beings.

Whereas we have been focusing on systems and institutions, this scenario takes the person as the starting point and focus of interest as both Chinese humanism and Buddhism do. In part, this is why we have chosen to call this scenario "Buddhist." The other principal reason is that we have been greatly inspired by E. F. Schumacher's development of an "Economics as if People Mattered," described in his noted work *Small Is Beautiful*. In this "Buddhist economics," which in his words is "the systematic study of how to obtain given ends with the minimum means," the goal then becomes to achieve the maximum well-being with as little physical throughput as possible. Peace and permanence are the watchwords of a Buddhist conception of economics and lead to a distinction between capital resources and income resources. Thus, to use nonrenewable resources "heedlessly or extravagantly is an act of violence," he states, and "squanders nature's capital, the nonrenewable resources" as if they were income. Profligate use of these resources also destroys a second kind of capital, "the tolerance margins of nature."

Many non-Western religions, including Buddhism, stress a relationship between man and nature which is one of harmony rather than hubris. In these religions, man is part of nature, not master of it, in striking contrast to the Judeo-Christian world view. Thus, when we say "Buddhist Scenario" or "Buddhist economics," we are not suggesting that we all become Buddhists. We use the term "Buddhist" rather as a symbol to evoke certain values which characterize Buddhism and many other cultural traditions and to refer to Schumacher's contributions to the present dialogue on ecological and economic issues. The next chapter will delineate more precisely the values which inform this scenario and describe briefly two societies which illustrate many CS_3 kinds of values.

We shall present three strategies which could facilitate the movement toward a CS_3 society. The kind of society we shall describe is clearly very different from the mainstream of present Western industrial society. Significant changes in our way of life would surely accompany a transition to a "Buddhist" kind of conserver society. We wish to emphasize, therefore, that if this society is to emerge it will do so gradually as people and conditions change to become more receptive to this kind of thinking and being. Significant changes are already occurring in this area.

CS_3 is the most conservationist scenario in its effects, and if a society chooses to take this path we suggest three principal strategies: NANG—Negative Artificial Needs Growth; NIG—Negative Industrial Growth (Deindustrialization); NUG—Negative Urban Growth.

If we accept the argument that the economy has become a want-creating or want-stimulating mechanism rather than just an institutional structure designed to satisfy existing wants, then we must accept a link between the growth of advertising and marketing since World War II in the United States and, for example, the following increases in industrial production: nonreturnable soda bottles (up 53,000 percent), synthetic fibers (up 5,980 percent), mercury used in chlorine production (up 3,930 percent), air-conditioner compressor units (up 2,850 percent), plastics (up 1,960 percent), electric housewares such as electric can openers (up 1,040 percent), synthetic organic chemicals (up 950 percent), aluminum (up 680 percent), electric power (up 530 percent).* Only a reduction—negative growth—in the stimulation of people's wants can bring us back to a saner level of material consumption, thus NANG.

Many have argued that this kind of industrial growth has brought with it not increased well-being, but rather inflation, ecological crises, shortages, social breakdown, personal alienation, pollution- and stress-induced diseases, and other ills. If we agree with this assessment, then NIG and NUG follow.

One philosophical argument, then, for CS$_3$ is that in focusing our society, our intelligence, our human energy, and our material resources on the development of a mass-consumption society, we have traded our higher selves for a higher standard of living in material terms. Our personal inner growth has withered as our external economic growth as a society has flourished. There is another equally important philosophical raison d'être for the creation of CS$_3$, the diminution of international inequalities. We stated in Chapter 6 that any conserver society must be compatible with the goal of reducing income inequalities, both nationally and internationally. Thus far we have not considered this international dimension explicitly. Let us do so now.

Implementing either conserver scenario one or two in the economically developed countries would have positive implications internationally vis-à-vis the developing world because efficient resource use by the heavily industrialized countries would free resources, material, financial, and organizational, which could then be devoted to the satisfaction of basic human needs not now being met. This argument gains even greater force if, as in CS$_3$, we explicitly take the individual human being as our point of departure. If we accept the Kantian maxim

*Barry Commoner, *The Closing Circle* (New York: Bantam Books, 1972), pp. 140–141.

that each human being by virtue of his or her existence is uniquely valuable then, logically, each person living in our world has "the right to a decent survival," as philosopher Franz Oppacher has expressed it.

Do the citizens of one country, the United States, who constitute 6 percent of the world's population, have the right to consume annually more than 29 times as much energy as the people of the Middle East, Latin America, Asia (not including Japan and the Soviet Union), and Africa; import 90 percent of the planet's tin and manganese; import 87 percent of the world's aluminum, 85 percent of the world's chromium and asbestos, 24 percent of the iron ore, and so forth; and feed 20 million tons of vegetable protein to animals which produce only 2 million tons of animal protein which is subsequently eaten? American-owned animals convert vegetable to animal protein at a conversion efficiency of only 10 percent; tens of millions of starving and malnourished people could be fed by the 18 million tons of protein lost in this process each year.

We could go on citing examples from other countries in the industrial world for many pages but the point is clear: while impoverishing ourselves spiritually in our staggeringly wasteful mass-consumption society, we are impoverishing the rest of the world materially. Our greed has been served at the expense of others' needs. This then is the second major philosophical reason for moving toward a conserver society, especially one like CS$_3$, wherein we consciously decrease our material consumption in order to increase our human satisfaction in our work, our families, and our communities. Being instead of buying becomes our watchword.

News for the Unenlightened from Rita Righteous

Rita was a typical flower child of the sixties. She had spent so much time supporting every conceivable good cause that she had taken a couple of extra years at college to finish her social-science courses. The anti-war movement; anti-nuclear-bomb demonstrations; committees to save the wolf, the seal, the whale, and the buffalo; civil-rights sit-ins; picketing against artificial preservatives in food and smoking in public places; combing back-street boutiques for clothes and furniture made anywhere but in North America had occupied her fully. If that was not enough, she had tried out every imaginable aspect of the human potential boom—yoga, transcendental meditation, positive thinking, primal therapy. Rita had earnestly pursued self-fulfillment wherever it beckoned. She had been to California, India, Tibet, and then, but not finally, back to Jesus.

Rita had met Sammy through Angus McThrift, who had helped her and nine other friends organize their cooperative in an old house near the university. She had been astonished and a bit guilty to find that, with Angus's "economies of scale," they could live almost in luxury while spending very little. She and Angus, being together a lot, had become involved with each other, but she had always had doubts that he could be a really good person while still "making money."

Sammy was nodding his head ruefully, thinking that it was inevitable that they would split up. They had entirely different attitudes to life. He hoped that Rita's letter would not be too regretful and depressing.

> Commune du Droit Chemin
> Ardennes
> July 18th

Dear Sammy,
    Please excuse the scribble but I'm writing this quickly between Chanting and the Community General Meeting. While chanting

I'm keeping you especially in mind to help you face the challenge of being forced to stay the whole summer in the city. It's tempting to tell you—although I'm new to it myself—about the wonderful things I'm learning here which would help you overcome all obstacles and regard even bad things with serentiy. At least you are suffering a bit—which is good. Angus seems to get everything his own way but he doesn't learn or increase his personal growth at all.

By the time we arrived in Europe I had realized that that's the basic difference between us and I left him at the airport. He was going to collect his new car –the symbol of all that's important to him in his life –flash, luxury. The fact that he is clever with money no longer impresses me. Greed and pollution are still his life – beating the system does not mean getting out of it. It's definitely finished between us now, as I have taken a very different path. I'll not go into all the details, but I met some people in Luxembourg and came here with them. We live very simply, working, sleeping, meditating, discussing philosophy and eating –our own beautiful, clean, organically grown food. It's a wrench at first to give up the unhealthy "luxuries" but it's getting easier every day. The dual principle of action and wisdom, informing each other, is always at work. As I'm sure you're learning, the bad karma of the past is difficult, but not impossible, to overcome.

Wishing you (and Angus, if you're writing) a new kind of happiness for the future.

Love,

Rita

17. "BUDDHIST" VALUES

In thinking about technology, we must distinguish between technology which enhances our potential and that which reduces us to servants. The right kind of technology, according to Schumacher, is characterized by smallness (human scale), simplicity, cheapness, and nonviolence: a technology which can be controlled and directed to serve our ends without overwhelming us in the process; a technology which serves people, not organizations. "Violence" refers not only to the tendency toward centralization and enormous size which do violence to the individual's unique personality in trying to cope with his or her problems, but also to man's relationship with nature.

The kind of technology which meets these criteria Schumacher has called "intermediate technology" (others have labeled it "low-impact technology" or "appropriate technology"). Most conserver technologies are of this type. Taking agriculture as an example, although many millions of farmers have been able to function well in the past without insecticides, herbicides, fungicides, and chemical fertilizers, we have adopted industrialized agriculture, with deleterious effects for the soil and the complexity of local ecosystems, and in the process done violence to nature in the name of "the Green Revolution." Yet we have not solved the problem of the world's food supply because these industrial methods destroy fertility and make people more dependent on costly industrial products, most of which are extracted from nonrenewable resources, oil and other hydrocarbons. Agriculture has become increasingly more industrialized in the size of its units, in its patterns of resource use, and in its emphasis on cash crops grown for consumption by the developed countries (coffee, cacao), which has not solved the central aspect of the world food problem: equitable distribution of the food we have raised.

A "Buddhist" kind of agricultural policy would emphasize local self-sufficiency and satisfaction of basic subsistence needs before entering into cash cropping. Sacrificing local control over one's food supply for a cash income, which is quickly diminished by inflation, has been the experience of many countries over the last few decades.

In many ways an agricultural situation wherein one becomes dependent and alienated parallels the industrial labor situation in the West, where work is frequently seen as a necessary evil, a cost or an input which employers try to minimize through automation and a sacrifice workers make in order to gain income. In contrast, according to Schumacher the Buddhist notion of "right livelihood" views work as fulfilling three functions: (1) to give us an opportunity to use and develop our capacities; (2) to help us overcome our ego-centeredness by engaging with others cooperatively in doing a common task; and (3) to create the goods and services needed for a becoming existence. This conception of work is based on a view of human needs which Schumacher has clearly articulated in his pamphlet, *People's Power*, which closely resembles the structure of needs elaborated by W. Lambert Gardiner and Franz Oppacher, a psychologist and a philosopher who worked on the Conserver Society research team.

Schumacher identifies three central needs: "to act as spiritual beings" or "to act in accordance with . . . moral impulses"; "to act as neighbors" or "to render service to their fellow men"; and "to act as persons, as autonomous centers of power," "to be creatively engaged, using and developing the gifts which a person has.''* If work satisfied these needs, then its meaning in Western societies would change dramatically and become intrinsically satisfying to people rather than just a source of income. People would be valued for their skills, their creativity, and their individual personal characteristics instead of according to their salaries or titles.

When work is done on a human scale in a cooperative manner, more with tools than with machines, a sense of community arises, with each contributing to the well-being of the group in a unique way. When large-scale hierarchical bureaucracies and complex machines are not present to inhibit the development of personal ties, intimacy and trust can replace role playing and anxiety in working relationships. Working does not have to be separated from living. One could work and play with one's family and friends instead of having two rigidly separated lives—the working life and really living (mostly on the weekend).

Thus, technology, work, and family and community life in our CS₃ scenario reinforce and sustain each other.

Two concrete examples of conserver societies from which we can learn much about these rich interactions and human interdependencies are the Cree Indians of Northern Canada and the Hutterian Brethren, who live in the American Midwest and in Canada's prairie provinces.†

The Cree and many other North American Indian groups exhibit a religiously based harmony which enables them to live on the land, peacefully hunting and trapping animals for food and furs without destroying them. Cooperation and sharing re-

---

*E. F. Schumacher, *People's Power* (London: National Council of Social Services, 1975), p. 3.

†The description of the Cree is based primarily upon one of the authors' fieldwork and publications (Peter S. Sindell), while the material on the Hutterites is drawn principally from John Bennett, *Northern Plainsmen* (Chicago: Aldine Publishing, 1969) and John A. Hostetler, *Hutterite Society* (Baltimore: Johns Hopkins University Press, 1974).

sources are central to those cultures as they are with Hutterites, thus maintaining a relatively equal distribution of such basic resources as food. The individual is devoted to the collective welfare rather than individual self-aggrandizement, as in the case of our societies. The Cree differs from the Hutterite in technology, but both groups remain conservationist in practice and collective in spirit, in contrast to our religion of mass consumption with its cathedral of the shopping center.

The Cree Indians

In order to understand the Cree Indian cultural tradition, a culture lived by tens of thousands of hunters, trappers, fishermen, and their families all across the Canadian boreal forest from Quebec to Alberta, one must begin with animals and man's sacred relationship with them. Animals and men are not sharply distinguished as they are in Euro-Canadian culture but are rather seen as different kinds of beings living together in harmony with mutual respect in a world alive with spiritual significance. A man is able to kill animals not only because he is an accurate marksman or a skilled trapper but, more importantly, because the animal *allows* itself to be killed. However, an animal allows itself to be killed only when the animal has been treated respectfully, for example, throwing the animal's bones in the water or putting them in trees after the animal has been eaten so that dogs cannot get at them. In this and other ways is the master of each animal species (e.g., the Great Moose or the Great Rabbit) made happy. The Cree maintain a good equilbrium with animals, using a "management system" based on the deep spiritual ties between men and animals.

Animals and man live together on the earth, "Our Mother." The Cree world is a unitary one of harmony and respect between the different kinds of beings: men, animals, and spirits. The distinction made by Whites such as "trapping" and "hunting" and "fishing" have no place here. They are imposed and artificial. What is crucial is the spiritual relationships between each hunter, his "power," and the world around him. One can see from this how devastating the idea of drowning the earth, slashing the forest, and disrupting and killing the animals must be to traditional Cree people. The Canadian James Bay and Churchill-Nelson River Diversion Hydroelectric Projects, emanating from the mass-consumption society, are destroying this Canadian conserver society.

Related to their view of the earth and animals are the Cree views of land use. Whereas Euro-Canadian law is based on the notion of ownership and, hence, allows destruction as a prerogative of the owner, the Cree believes that everyone is free to use the land and its fruits but not to destroy it. In the area of land use, the Cree display a cooperative rental model of the kind described in CS_1.

Since animals, birds, fish, berries, etc., are scattered lightly over the landscape in the boreal forest, traditionally small hunting-trapping-fishing groups consisting of ten to twenty-five closely related people would spend their winters exploiting large territories, which usually were passed along within particular extended families. During the summer, when fish or other resources were plentiful (again, a concept of European origin—"resources"), several of these kin-based subsistence groups would gather to rest from the arduous winter, make marriages, repair equipment, and prepare berries, fish, and other foods for the coming winter. With the development of the fur-trading posts, these gathering places, which were always rich in resources, became more permanent communities. Nevertheless, in most isolated northern Cree settlements, until compulsory Euro-Canadian-style schooling was imposed, relatively few people stayed in the settlements during the whole winter. Even during the summer, when people had returned from their hunting grounds, at any one time many people would be away from the post fishing, hunting, or berry picking.

Related to the issues of leadership and authority are Cree values relating to individual autonomy, self-reliance, and the collectivity. Because of the harsh climate, the dangers of bush life, and the subsistence pattern, development of individual autonomy and self-reliance are necessary for survival. But individual autonomy and self-reliance are not used for the aggrandizement of the individual as they would be in our consumption-oriented society; rather they are placed at the service of the group, normally the kin-based hunting-trapping-fishing group. Concomitant with this emphasis on the needs of the group is a strong emphasis on generosity and sharing of both food and labor. Cooperation, not competition, is the hallmark of this conserving society and is reflected throughout.

Men and women are esteemed not for the accumulation of goods, but for their wisdom, their spiritual power, their skill in hunting, and their generosity in sharing with others. Euro-Canadian society operates economically on the principle of balanced reciprocity—I give you X and you give me Y, which is of equal value—while Cree society operates on the principle of generalized reciprocity—I give you X and somebody else gives me Z and you may give me G, and so forth, with no expectation of a specific exchange even in the future. In some cases, for example, old people who have no children or grandchildren to help them, it is known that one's gifts probably will never be reciprocated, but that is no problem because one gives because one (autonomously) wants to, not from any ulterior motive.

The Cree world view and values are transmitted in the context of a society that is small in scale, personalistic, and kin-based, quite the opposite of Western mass society. Culturally, spiritual power and generosity are valued, rather than prestige based on material goods or cash income.

The Hutterite Brethren

Self-imposed isolation, an agricultural economy, and a communal style of life characterize the religiously-based Hutterian communities of the three Canadian provinces. The rigidly Christian Hutterite doctrine defines mass-consumption society as corrupt and evil because of its pervasive pursuit of money, possessions, and selfish personal gratification. Their agricultural economy allows them the opportunity to reject urban, industrial society. Communal living intensifies their isolation; they have their own system of government, and social interaction is almost exclusively within the Hutterian circle. The cultural maintenance of Hutterite communities, then, necessitates a great degree of internal economic efficiency in order to preserve their isolation and thereby carry on the faith. Despite their economic success, austerity is the central ideal of their religion.

This difficult task, joining great economic efficiency and success with an individual austerity, depends on conservationist use of land and other resources and economies of scale. The Hutterites represent an excellent model for the study of an efficiency-oriented yet conserving society, which at the same time places spiritual values first in importance.

The population of a Hutterite colony is intentionally kept small, allowing definite and purposeful roles for all members of the community. When a colony grows so large that this type of cohesion begins to wane, then it divides into two colonies. Thus, every colony has the incentive to work hard and amass capital through savings to support the inevitable future offshoot colony. Colony division also has a modernizing effect, since new colonies commence with a small population and have a limited labor supply. They need, therefore, to use more and newer labor-saving equipment than the larger colony.

While a future-oriented outlook may be economically and socially required as the mother colony prepares to spawn a new colony, future orientation also plays an integral part in Hutterite spirituality. Earning the grace of God in the life after death is the central goal of Hutterite life. Only man's spiritual nature is considered to be good and valuable. Thus, earthly work is done only to maintain them physically until they enter their truly satisfying afterlives.

John Bennett characterizes the Hutterite way of life as "living with less and doing so with dignity and purpose." Hutterite colonies are communal in all aspects of the economy: production, consumption, and distribution. Property is defined as the right to use but not to possess, which certainly affects their concept of basic needs. These minimal needs are defined by consensus and are generally provided for by the community. For example, married women receive a specified yearly allotment of cloth to make their family's clothes and a newly married man will be given certain basic pieces of furniture for his household's use. In the distribution of goods, impartiality and sharing are emphasized.

The overall efficiency of the Hutterites' communal society is based on cooperation. Cooperation exists first on the basis of a "managed democracy," a combination of egalitarian group decision making and much-respected patriarchal authority. It exists secondly in the reciprocal nature of the Hutterite system, in which goods and profits are for the welfare of the group. Each member contributes labor, time, energy, and earnings to the collective and in return the community supports, instructs, and educates every individual member of it. An important aspect of the cooperative Hutterian society is its internal diversity, which provides year-round work for everyone. While cooperation may enhance efficiency and productivity, colonies have learned that cooperation and diversity with small total profit margins are often better for the group as compared with a few enterprises with large profit margins. One would assume that cooperation and group cohesion are enhanced by this diversity of functions and the continuous availability of productive work.

Cooperation in Hutterite society is also expressed in the col-

lective use of goods and skills. Within single colonies and between different colonies, Hutterites use and exchange machinery, labor, and knowledge for the advancement of all. For
example, there is generally only one passenger vehicle for the
use of the whole community, usually consisting of thirteen nuclear families. Cooperation in labor within a colony even ex-
tends to people helping others with their unfinished work.

This consumptive austerity or "consumption control,"
coupled with an extremely high productivity relative to the resources used, affords the Hutterites a large amount of investment capital, which then is used to reinforce efficiency through
the acquisition of better and better machinery. A large amount
of investment capital in this "economy of scale" therefore actually permits greater conservation in the use of resources, since
the Hutterites are able to take bad land out of production for a
few years and convert it to other uses. Most individual prairie
farmers cannot afford to do this because they lack sufficient
capital.

Efficiency is further seen in the tendency toward recycling.
The mechanical shops of Hutterite communities often utilize
scrap iron and steel from city dumps. They also buy used
equipment and fix it or adapt it to their own needs. Furthermore,
they repair electric motors and generators secured from discarded equipment. Ingenuity in this regard also appears in the
colonies' use of large discarded petroleum tanks as grain storage
bins: high efficiency at minimum cost using recycled objects.

The fact that the Hutterites are able to support larger populations on the available resources than their individual farming
and ranching counterparts speaks for the success of their system. Colonies, in fact, do not draw welfare, old age, or unemployment pensions, and most do not even accept the family
allowances due them. In terms of size, Manitoba Hutterite colonies are more productive in every major agricultural enterprise
except cattle feeding. This is achieved on approximately half of

the acreage used by comparable non-Hutterite farm families in the province and generally with a more conservationist approach to the land.

Some might say that this discussion of "Buddhist" values is all very well but has little to do with our real world, which is urban and highly industrialized. But we argue on the basis of ethnographic observation (since no statistically valid scientific survey data exist) that a significant and growing number of people of Canada, especially in the younger age bracket, are moving away from the materialistic, consumption-oriented values promulgated by industry and commerce through advertising, and that some are rejecting life in the large urban centers. The evidence which does exist suggests very often it is the children of the urban middle class who are participating in this movement. There seem to be two themes: self-realization and self-development; and resurgence of emotionally warm, communally-oriented relationships. These trends are sometimes intertwined; in other instances the only sign of value change may be the development of a new social activity within an office or a group of friends, such as swapping cuttings from their house plants.

Our attitude toward nature is changing, as can be seen in the rapid development of health-food stores and their products. Rejecting "plastic food," which has been processed with chemicals, bred genetically for size and harvesting characteristics, rather than taste and nutrition, and packaged extensively, constitutes a significant and visible rejection of industrial mass production and marketing of goods. The return to the methods of the old-time general store, which characterizes most of the health-food movement so far, has enormous implications for conservation in terms of packaging and the distribution economies possible when products are displayed and sold in bulk. The consumer is usually required to bring his or her own container or bag for the grains, nuts, dried fruits, and so on. More intimate social interaction also occurs in the health-food store as the proprietor and the customer chat while measuring and pricing the goods—another contrast to the impersonality of the large supermarkets owned by the major chains. Using the shelves of book stores as a sensitive indicator of current tastes and trends, we can observe a tremendous increase in the number of books about health, organic and wild foods, herbs and spices, as well as plants, while the growth of communally-shared gardens, roof gardens, and window boxes also supports the notion that this trend reflects a new concern with nature as well as a conserver-oriented kind of consumption. Plants have durability, are never obsolete, and constantly grow and reproduce themselves, while fruits and vegetables grown at home lead to some self-sufficiency and save packaging and distribution costs.

The growth of religious groups which abjure egocentric approaches to self-realization and discourage material consumption also seems to be a significant and socially visible trend. "Losing oneself in order to find oneself" is returning as a spiritual route to self-fulfillment as an alternative to the extrinsic orientation of materialism. Yoga, transcendental meditation, sensitivity training, encounter groups, and other aspects of the so-called "human potential" boom do not constitute new religions but again represent a shift toward spiritual ways to achieve self-actualization. It is significant that many of these movements have biocentric orientation in contrast to an orientation toward mastery over nature.

Investigators in the United States have recently detected and described social trends which they call "Voluntary Simplicity." Duane Elgin and Arnold Mitchell* define "Voluntary Simplicity" as "living in a way that is outwardly simple and inwardly rich. This way of life embraces frugality of consumption, a

---

*"Voluntary Simplicity (3)," *Co-Evolution Quarterly* (Summer 1977): 5-19.

strong sense of environmental urgency, a desire to return to living and working environments of a more human scale, and an intention to realize our higher human potential, both psychological and spiritual, in community with others."

They cite a Harris survey released in May 1977 that concludes by stating:

Taken together, the majority view expressed by the cross section of 1,502 adults in this Harris Survey suggest that a quiet revolution may be taking place in our national values and aspirations. Some of these attitudes reflect the energy crunch and the realization that the supply of raw materials is not boundless; others are a legacy of all those ideas that young people pressed for in the 1960's that have now begun to take root in the 1970's.

Whether the American people are prepared to face the consequences if the country follows the choices they so clearly express is quite another matter. But there is no doubt that there has been a profound shift in many of the traditional assumptions which have governed the nation.

On specific subjects the poll showed the following:

82 percent would concentrate on "improving those travel modes we already have" rather than "developing ways to get more places faster."

79 percent would place greater emphasis on "teaching people how to live more with basic essentials" than on "reaching higher standards of living."

77 percent come down for "spending more time getting to know each other better as human beings" instead of "improving and speeding up our ability to communicate with each other through better technology."

76 percent opt for "learning to get our pleasure out of non-material experiences" rather than on "satisfying our needs for more goods and services."

59 percent would stress "putting real effort into avoiding

doing those things that cause pollution" over "finding ways to clean up the environment as the economy expands."

63 percent feel that the country would be better served if emphasis were put on "learning to appreciate human values more than material values," rather than "finding ways to create more jobs for producing more goods."

66 percent would choose "breaking up big things and getting back to more humanized living" over "developing bigger and more efficient ways of doing things."

64 percent feel that "finding more inner and personal rewards from the work people do" is more important than is "increasing the productivity of our work force."

59 percent feel that inflation can better be controlled by "buying much less of those products short in supply and high in price" than by "producing more goods to satisfy demand."

While polls are surely not conclusive indicators of social change, these results must give pause to those who see CS₃ as utterly impractical and utopian.

Although their projections are not founded on hard data because this area of research is so new, it is striking that Elgin and Mitchell suggest that there are at present 5 million Americans fully involved and 10 million partially involved with what they define as "Voluntary Simplicity Tenets." They project a maximum possible growth by A.D. 2000 of 60 million full Voluntary Simplicity people and 60 million partial Voluntary Simplicity people, with 25 million as sympathizers, and 55 million as indifferent or opposed.

We suggest that many of the values we have described as characteristic of CS₃ are gaining more and more acceptance in Western industrial societies, especially among the young, well-educated, and affluent who have found little satisfaction in ever higher levels of material affluence.

The link between changing our own life-styles in a search for more meaning and satisfaction, while consciously cutting down our own consumption levels to benefit others who have far less, is explicitly and powerfully presented by the Simple Living Collective in their excellent book *Taking Charge*.* Everything from an energy addict's calorie counter to discussions of food, clothing, health, community, children, work, personal growth, creative simplicity, and the linkage between personal change and political power, nationally and internationally, are presented. *Taking Charge* is a handbook for raising our consciousness and implementing in our daily lives many of the facets of CS_3.

---

*N.Y.: Bantam Books, 1977. Another handbook has just appeared along similar lines called *Progress As If Survival Mattered: A Handbook for a Conserver Society* (San Francisco: Friends of the Earth).

18. NANG

or Negative Artificial Needs Growth

Our central goal in this third scenario is to achieve the highest possible degree of human satisfaction or happiness for each person in the society with the lowest possible amount of physical throughput. "Doing Less with Less" means doing less materially while doing more to fulfill each person's higher needs. But what are these higher needs? Doesn't our society satisfy them now? What kind of person would be happy in this kind of conserver society?

All of us share an evolutionary history of several million years. Our nervous systems are similar in the way they are structured and the way they work. On a basic level we have the same needs. Where we differ from each other, both from culture to culture and from individual to individual, is in what we want to satisfy these needs. For example, we all need food, but in traditional Chinese culture rice is thought of as the best food whereas in Latin America many people would choose tortillas made from corn flour. Most Americans and Canadians, given a choice between rice, tortillas, and steak, would choose steak. In every situation, people are satisfying a requirement for food, but what they choose to satisfy that need depends on where they grew up and what they learned to like and to think was a good meal. We can call these culturally based need satisfiers "artificial needs" or "wants," in contrast to the real need for food.

Many psychologists, anthropologists, philosophers, and psychiatrists rank needs in terms of their importance to us for our survival and happiness. Franz Oppacher, the philosopher in the Conserver Society research team, identified needs in seven categories, all of which are very much interconnected, of course:

1. Physiological maintenance: related to physical health and the quality of the environment.
2. Physical security: maintenance of the physical structure of the organism, i.e., survival without injury.
3. Psychological security: maintenance of psychological or personality structure. Many of the needs in this category are related to the demands for a sense of identity, a feeling of control over one's destiny, and stability over time in one's understanding of the world, which we can call in summary "cognitive control," following anthropologist George D. Spindler.* This set of needs is closely linked with what W. Lambert Gardiner, the psychologist on our research team, has called the needs for stimulation and consistency.
4. Love, belonging, cooperation: comprising mutual (including parental) love, physical love, comforting, a sense of community, and emotional relationships with people and places. Alienation is a symptom signaling lack of fulfillment of this need.
5. Self-respect, personal dignity: including the sense that one's goals are worthwhile, one's work is useful, and one's efforts, skills, and personality are appreciated by others. This need set also includes the desire for self-esteem or a positive sense of self.
6. Self-actualization, growth, competence: related to exploration, play, curiosity, creativity, aesthetics, and allied to Gardiner's need for "knowing" one's environment.

* "Psychocultural Adaptation," in *Personality and Culture: An Interdisciplinary Appraisal*, E. Norbeck et al., eds. (New York: Holt, Rinehart and Winston, 1968).

7.   Understanding, purpose: Frankl's "will to meaning"*— close to Gardiner's need to "understand" one's environment; this also includes the need for religion.

These seven categories of needs are tentative and probably incomplete, but they can serve as a first step in helping us consider all of our various needs. They also guide us in thinking about our present mass-consumption society and whether this society helps or hinders us in fulfilling our needs. At the lowest level we have biological needs which relate to our survival as organisms. Next, we have social and psychological needs, which are just as necessary as food or sex for our healthy development as human beings. At the socio-psychological level, for example, if we don't get enough caring and intimacy as children, we cannot become normal adults and may even die.

Many critics of the mass-consumption society would argue that we have created a society in which we try to satisfy all of our needs, high and low, with *things*. If we want love, we are told we will find it only if we use brand X deodorant or fly to exotic islands with Fantasy Airlines—twenty months to pay and ten dollars down.

If the gross national product is our god, then it follows that our main purposes in life are to produce goods and services and then consume them, on credit if we can, to stimulate the economy even further. Linked with this mathematical image of life and its meaning is a view of the human personality which we call the "consumer personality."

W. Lambert Gardiner is fond of saying, "I am compensated for my life with money and then I vainly try to buy it back."

Until recently, many psychologists and most economists have had a behaviorist image of human nature. With this kind of simplistic "stimulus and response" thinking, people become mere responders, the nervous system is there just to act as a

* Viktor Frankl, *The Will to Meaning* (New York: New American Library, 1969).

go-between (to mediate, in technical terms) between the organism and his or her environment. Needs then are extrinsic or external to the nervous system, which just responds to, say, the need for food.

In this view of human beings, everybody seeks instant gratification—nobody has any sales resistance. Small immediate gains are preferred to larger, but remote, joys. Thus every person is seen as a consumer, an insatiable consumer at that.

Logically, if one responds only to the external world and one's internal biological needs, there is little scope for "free will," "mind over matter," or "self-actualization." There is no place for an intrinsic sense of worth without these and, thus, little sound basis for self-esteem or a positive self-image. If our self-image is based on money or clothes, a big house or a color television set, without these what are we? who are we?

It is a frightening thought that if our principal value to society is as a consumer and we act as consumer personalities, then our relationships with others in society take on this cast.

As we sit alone in front of the television set (even if others are there), our family relationships and feelings wither. How can deep feelings for others develop while we are shopping or watching TV? This too is an aspect of the consumer personality, a lack of intimacy, a quality of alienation. Contractual relationships between people who are playing roles instead of interacting as people per se become the rule. Intimacy cannot come, Gardiner argues, from people whose motivation is extrinsic, outside of themselves. Family and community as emotionally satisfying networks of love, friendship, and affection are diminished by a preoccupation with things.

Making, promoting, advertising, and marketing have come to dominate our individual and collective lives as the mass manufacturers and merchandisers seek to create even more artificial needs while telling us each time that the new product will lead us to happiness.

Taking a recent example, to introduce one new "natural" cigarette, $50 million, 130 boxcar-loads of advertising materials, and 25 million sample giveaway packages were thrown into the breach. In contrast, each year the U.S. government spends less than $2 million warning American citizens of the health hazards caused by cigarettes, which directly lead to the death of about 250,000 Americans each year. The advertising executive in charge of the new cigarette said, "Before long you won't be able to turn around out there without having [brand name] hit you over the head."*

Along the same lines, there are massive campaigns to sell powdered milk in the Third World by trying to convince mothers than their natural milk is not as good as the manufactured product. Some scientists have suggested that a rise in infant death in many Third World countries is directly linked to this phenomenon, since the water used to prepare the milk is often contaminated, leading to diarrheal disease and death.†

We ignore life and promote death when we have a view of human nature which is limited to material consumption. Instead of optimizing material consumption, that is, choosing the right level for it, we have maximized it—bigger is better, no matter what the consequences.

A conserver personality, in contrast, is based on a recognition that all needs must be in harmony: biological, social, and psychological. The view of the self is humanistic and needs for intimacy, creativity, autonomy, insight, and stimulation are recognized and given the importance they deserve.

The whole person with his or her rich diversity of needs and potentials is present in this conception of people's psychological makeup rather than the narrow unidimensionality of the consumer personality. It follows then that conserver personalities can have intimate relationships interacting as one person to another instead of clerk-to-customer, doctor-to-patient,

*The New Yorker, June 27, 1977.
†New York Times, April 6, 1976.

worker-to-boss, and so forth. Relationships involving our whole beings are possible because we act and interact on a basis of intrinsic worth and psychological integrity, the kind of worth which is possible only when one controls and directs one's own life as opposed to merely responding to organic needs and external stimulation.

Needs in a conserver personality reflect organic *potentialities* instead of organic requirements. The nervous system does not only mediate. Gardiner cites higher-order psychologic needs for stimulation and consistency which provides the organic base for actively wanting to know our environment (assimilation of information) and to understand it (placing this new information into coherent subjective maps of our worlds). We are not just passive consumers, either of goods or of information; we seek and strive and love and hate and feel in autonomous, independent ways.

If we accept this view of ourselves and our human potential, the kinds of people we are, the meaning of work, the way we see ourselves vis-à-vis our tools and technologies, all these change toward the Buddhist notions of personhood, work, and the proper role of technology. We cease to be just automatons of the marketplace. We refuse to be "fitted into jobs" and defined only in terms of what cog we are in the bureaucratic or industrial machine. With this new *optique* we also make sure technology serves us instead of being servants to its purposes. How can we help people move in this direction?

Our central suggestion here is that if people can satisfy their higher-order needs better then they will no longer want to trade their higher selves for a higher level of material consumption. At the same time, the creation and direct stimulation of artificial needs or wants which require high throughput must decrease significantly.

In terms of Buddhist values, fulfillment of needs for intellectual stimulation, personal growth, interpersonal intimacy, individual autonomy, and spiritual self-realization will increase when decentralized, cooperative, small-scale, and relatively self-sufficient communities emerge. To encourage the development of such communities, whether they are based in urban neighborhoods, rural villages and towns, or suburban housing developments, we would emphasize local control over decision making and problem solving. E. F. Schumacher has remarked that governments are good at collecting money but people are much better than bureaucracies at spending it. Consequently, grassroots programs which arise from felt community needs should be expanded. Examples include the Local Initiatives Programmes and Opportunities for Youth in Canada, and in the United States many of the ideas developed in the sixties by the Office of Economic Opportunity, for example VISTA.

Logically voluntary associations and community-based institutions would be funded to replace governmental bureaucracies wherever possible. Along the same lines, support for theaters, museums, musical groups, artists of all kinds would expand dramatically. Noncommercial community festivals, arts and crafts shows, recycling centers, and community-owned and operated radio stations, television channels, newspapers, and magazines would become priorities. Cooperatives of all kinds—industrial, farm, and consumer—would be assisted in their development. The ratio of profit-making to nonprofit institutions would change while both kinds of organization would be retained. Educational, cultural, and recreational centers and associations based on geographical proximity, language, culture, or shared interests would flower if resources were allocated more at the local level. As more local self-sufficiency in agriculture and industry developed, based on appropriate technology using principally renewable and recyclable resources, funds now spent on maintaining enormous bureaucratic and technological infrastructures could be rechanneled for local use. Savings in the transportation and energy sectors alone, derived from a shift toward more local production of food, energy, and consumer goods, would be substantial.

Municipal sewage-treatment costs and water use would decline extraordinarily with the introduction of the clivus multrum. This is a Swedish toilet that uses no water and produces a sterile organic fertilizer from human wastes through bacterial action over a period of months. The humus produced is excellent for gardening or agriculture, and the toilet itself has proven its efficiency for generations.

Another argument for decentralization and small scale in technology of course is that the present massive scale of our systems leaves us extremely vulnerable to disruption and disaster. In this context, brownouts and blackouts take their place with contaminated water systems, dangerous consumer products, and polluted food. If the water in any large urban center were to become a danger to drink or use, the consequences are almost unimaginable. Terrorists' disruption of power plants (whether nuclear or non-nuclear) and subway systems are in the same category, as are strikes in the vital public-service sectors: garbage collection, hospitals, police, and fire departments.

Decentralized technology and local decision-making responsibilities could obviate many of these hazards. The alienation so common now in our large urbanized mass-consumption societies would presumably decrease as people began again to take control over their own lives. As the size of our institutions decreased, citizens, employees, neighbors—all of us in our various roles—would be able to participate more actively in shaping our destinies and our daily lives.

Concurrently, in terms of developing our human potential, all who want to work would be allowed to, regardless of age. The "human obsolescence" which characterizes our present society (lay-offs, unemployment, forced retirement) would have no place in the Cs_3 Buddhist kind of society. A right to a certain amount of access to education resources during one's lifetime would also be established, as Ivan Illich suggested with his educredit-card scheme. As part of this development and in accordance with our emphasis on decentralization, many local skill-exchange centers and free universities would be founded. In Toronto, New York, California, and elsewhere, such new social forms are already springing up, even without formal support. In many such groups, each person devotes a certain number of hours to doing something, for example plumbing or teaching French, and, in return, can have the services of someone else in the group for the same number of hours. All work, whether done by a nuclear physicist for a citizens' probe into nuclear energy or a potter teaching his or her craft to an aspirant potter, is considered equivalent.

If the strategies we have described were implemented, people would become more interested in fulfilling their basic human, emotional, physical, psychological, and intellectual needs than in trying to achieve satisfaction of these through ever more objects of consumption.

Nevertheless, since our present market mechanisms have become want-creating rather than just want-satisfying mechanisms, the other side of NANG must be to decrease or eliminate the artificial stimulation of wants. This would require banning all advertising which the consumer is forced to watch or listen to because it is integrated into the regular programming. In some European countries, advertising appears for a set period each day so that the listener or viewer can choose whether to attend to it or not. Alternatively, a channel devoted only to advertising could be created. Banning unsolicited promotional mailings and giveaways by commercial organizations would also reduce the pressure on people to consume things for which they felt no need until exposed to the psychologically sophisticated blandishments of the mass merchandisers. Establishment of credit accounts without prior requests as well as advertising designed to stimulate purchases on credit or create other kinds of consumption debts (bank loans, finance company loans), should also be prohibited. Not allowing advertising on packages and preventing packaging not necessary for preservation or health purposes are other ways to reduce the stimulation-of-demand throughput.

All advertising could also be regulated to ensure that only purely informational content is presented. If such regulations existed, advertising which used psychological tricks to arouse false associations and expectations or false fears would clearly be prevented. To assist purchasers further to satisfy their *felt* needs, consumer information centers could be created where one could find out the range of products available, their costs and characteristics, availability, and so on. And advertising and promotional literature could include, of course, price data, distributors' names and addresses, the product's virtues and advantages compared to other products, cost, and similar content useful to the purchaser.

As people become more satisfied in other ways, a decrease in material consumption should result. Concurrently eliminating or significantly decreasing the intentional stimulation of new wants will significantly reduce throughput. Applying these two principal strategies should lead to NANG: Negative Artificial Needs Growth.

If the fulfillment of people's higher needs is more valuable than an ever higher level of material wants and their satisfiers, then CS_3 offers far more to us by doing less with less and by doing something else.

19. NIG and NUG

Inspired by the preeminent importance which we attach to harmony with nature, the goals which emerge to guide our actions in a CS_3 society must include minimizing our disruption of natural ecological processes and systems, attempting to achieve a stable rather than a deteriorating relationship with our environment, and maximum conservation of energy and materials, especially nonrenewables. Naturally we must also adopt the principles and values appropriate to Buddhist economics, which means an economy based on permanence of stock rather than flow. We must concern ourselves with the quality of air we breathe and the cleanliness of our water rather than with growth in air-conditioning sales and industries which supply equipment for cleaning up polluted water. In short, we must replace the gross national product as a measure of our societal well-being with quality-of-life (QOL) measures. As Franz Oppacher has remarked, QOL is to a conserver society what GNP is to a consumer society.

In this light, there are two further strategies for encouraging the development of a CS_3-type of society: NIG: Negative Industrial Growth and NUG: Negative Urban Growth.

NIG

Edward Goldsmith and his fellow editors of *The Ecologist* argue persuasively that humankind has been ill-served by the industrial urban society we have created. They have proposed a blueprint for survival, a plan for deindustrialization, deurbanization, and cessation of the intentional proliferation of wants. The conceptual underpinning of Goldsmith's argument is that it is "the biosphere, in fact—the real world—which is being industrialized," and in so doing a "new organization of matter is building up: the technosphere or world of material goods and technological devices; or the surrogate world."*

The crucial point, of course, is that the surrogate world can be created only by making use of resources extracted from the biosphere or the real world. This extraction of resources from the real world leads to its deterioration and contraction, and to substitution of a surrogate world of factories, cities, motorways, and airports, which then give rise to waste products. Because the processes of our industrialized surrogate world are much less complex than those of the natural world, they give rise to many more wastes. These then accumulate in an effectively random manner.

Goldsmith has pointed out the importance of the second definition of entropy, the destruction of organization. By destroying the organization of the real world and replacing it with a simplistic man-made world, we have effectively denied the needs of the system as a whole for stability and self-regulation in such a way that industrial society satisfies its own incestuous needs rather than those of the whole. Yet, since we are a part of nature as biological organisms, we are destined to suffer directly from the destruction of the world ecological system. We foul our own nest, so to speak. This transformation of the biosphere into the technosphere, Goldsmith remarks, has been called "development" and "progress."

If we are to achieve what Goldsmith and his colleagues call a "stable society," we must establish a stable relationship to the environment, minimize use of scarce resources, decentralize

---

*"A Plan for Deindustrialization." Presented to the Houston Conference on Limits to Growth, November, 1975, Mimeographed, p. 3.

industry, foster product durability, decentralize agricultural production and reduce its scale, and reduce industrial scale.

In establishing a stable relationship with our environment, we must substitute natural means for technological means in agriculture and industry. For example, "biorational" technology, such as species-specific insect growth regulators and pheromones, would replace nonspecific chemical pesticides as organic fertilizers (derived from human, animal, and plant wastes) would replace chemical fertilizers. Monocultural agriculture on an industrial scale would give way to the "old-fashioned" but efficient principles of crop variety and alternation of species in cultivation. All of the initiatives in CS_1 and CS_2 would be continued, as well as requiring labels on every consumer product to suggest the best ways to recycle and/or biodegrade them.

To minimize the use of scarce materials, instead of giving a depletion allowance, taxes on new materials would be related to availability. The scarcer a resource is the higher the tax would be on its use. Instead of giving depreciation tax credits for machinery and capital investments, production which maximized labor and minimized physical throughput would be favored. This could be done by removing all sales taxes on 100 percent handcrafted goods. Handicraft production and cottage industries could be given tax breaks. This notion fits well with the Buddhist emphasis on the joy of work and the ideal that work and life should be harmonious and integrated, not alienated. Naturally, in CS_3, policies which would assist in humanizing the work place would develop as mentioned previously in discussing cooperatives, worker participation, and profit-sharing.

In CS_3, in addition to all the policies implied by the CS_1 and CS_2 scenarios, one could argue for the addition of outright subsidies for plants producing and using fuels derived from organic wastes, such as methanol (derived from paper, wood, and other organic wastes). Decentralization of industry could be promoted through the revitalization of railroad services to rural areas and through tax credits given to small industries located in thinly-populated, nonmetropolitan areas.

Again using the tax system, taxes to foster product durability, energy conservation, and agriculture could be introduced. On this latter point, taxing land on the basis of its actual agricultural use rather than its potential for nonagricultural uses, loaning start-up funds to small freeholders, and assisting in the development of local farmers' markets would all help to develop small-scale agriculture oriented toward local self-sufficiency.

NUG

Many of the policies to support decentralization of industry and agriculture would provide the economic basis for increasing rural population. Improving the quality of life in rural areas as well as increasing the quality, frequency, and availability of rail and bus services between urban and nonurban areas would support the already increasing trend toward deurbanization. Tax credits to those who don't generate as much throughput as urban dwellers would also help to decrease the population of urban areas, as would giving loans and grants to start farms, businesses, and professional services in nonmetropolitan areas.

Conserving wilderness areas by respecting the lifeways of hunters, trappers, and farmers in the Arctic and in tropical rain forests would also curb urban growth. As urban dwellers, we are in the habit of exporting our problems to those areas instead of living in harmony and stability with our own environments. We dump our wastes in the countryside; we export the problems created by our wasteful use of energy to our rural and northern areas by building ever more massive hydroelectric dams, nuclear power plants, and oil and gas pipelines.

V

THE SQUANDER SOCIETY

or Conserver Society Minus One

20. IS THERE A CASE FOR THE SQUANDER SOCIETY?

There is a board game that has been available in stores since the early seventies called Squander. It is based on the same principles as Monopoly, and the idea is to spend a million dollars as quickly as possible, the winner being the most successful squanderer. This game is not unlike the extreme squander society, not quite the status quo but the direction towards which it could move.

The value system of the squander society can be built on three postulates which negate the three assumptions of the conserver society. Specifically, we would have to believe that: (a) there is nothing particularly wrong with waste and in fact it is a good thing; (b) the environment is both a golden goose and a garbage dump and we may not be using resources fast enough; and (c) the only time period worthy of consideration is now, the immediate present, and damn the future (and futurologists).

The first postulate could conceivably be defended. Waste is a subjective notion anyway. An individual (or group or nation) has the ultimate right to dispose of what he or she owns, including waste. Waste becomes the final expression of individual sovereignty and is a reflection of high freedom. Even if waste were an evil, it could be argued, it would be a lesser evil than the loss of freedom involved in attempting to eliminate it. Therefore, on libertarian grounds, one should be allowed to waste.

The rebuttal to this argument is just as easily constructed. Perhaps waste is indeed an attribute of freedom, but in a democracy the governing rule is that the freedom of any one individual stops where that of his or her neighbor begins. To forbid me to harass my neighbor, if I feel like it, is certainly a constraint on my freedom. However, to allow me to harass him is a constraint on his. Wasteful practices rob both our contemporary "neighbors" and future unborn generations of *their* freedom by effectively foreclosing certain options which would otherwise be available to them. Therefore, it is precisely because of the democratic principle of freedom that waste should be forbidden, especially in an emerging era of scarcity.

The second postulate, that environmental problems should be ignored, is attractive to those who think they can escape the consequences of their own acts. At the individual level, we may be tempted to unload our garbage on a distant street or deserted road, and generally pollute the environment, as long as we think we can escape that locale. This was still possible in the early period of industrialization, when the "fruits of the factory," so to speak, were enjoyed by only a select few. But in our mass-consumption society we are all in the same boat. On this "Spaceship Earth," and in its "compartment North America," we now have masses of consumers and producers. We can no longer afford to lay waste our host, the biosphere. The beaches of the Riviera were not dangerously polluted when La Promenade des Anglais at Nice was enjoyed as a winter resort by only the European *haute bourgeoisie* in the nineteenth century. Today the beaches of Europe are polluted because we have mass consumption and our Great Lakes are polluted because of our mass-production society. Our atmosphere receives the effluent from millions of cars, not just from a select few. Our cities produce the garbage, not just of a handful of consumers but of the entire urbanized population, the upper, middle, and lower classes. We therefore cannot ignore our environment with impunity because we are operating in an increasingly cramped compartment of an increasingly overcrowded spaceship.

The third postulate of a squander society favors a very short-term outlook. We would have a "now" society, in which all satisfactions would be demanded right away. If, in fact, the end of the world were just round the corner, then conservation would be futile in the same way that a human being, in the final throes of a terminal disease, would be ill advised to buy retirement-savings plans. But, unless the Apocalypse is around the corner, conservation is a better idea.

The case for squandering is, on the whole, rarely made in terms of our three postulates but instead is implicitly made in connection with one of the sacred cows of our times: creating jobs. Under cover of this magic phrase, a thousand and one outlandish schemes are launched by government agencies all over the Western world to provide employment. That these programs are more often than not a sheer waste of public funds is usually forgotten as long as the ritual of pretending to keep people employed is observed.

Conversely, all conservation, economy, and efficiency measures are attacked because in many instances they reduce employment needs—for the very simple reason that they economize labor. The paradoxical situation is that, whereas we applaud a saving of oil, resources, and materials, we are chagrined by a saving of our human resource labor and we react by deploring the "unemployment" it creates.

In an attempt to clarify the issues, first let us advance the following proposition: *The more wasteful a society the greater the employment opportunities.*

Assume half the production coming out of the factory would immediately self-destruct. The result would be an unsatisfied demand for the destroyed half with increased employment opportunities to reproduce it. In essence this is what happens in wartime, a period, by the way, rarely accompanied by unemployment. Goods are produced, they then go up in smoke on the battlefield, and more goods are produced. Meanwhile the labor force is absorbed either as cannon fodder or as producers of the goods which are scheduled to go up in smoke. Even those normally outside of the labor force are called upon to do factory work. When the war ends, planners panic because they expect serious unemployment. At the conclusions of the Second World War, the Korean War, and the Vietnam War, the danger of economic disruption loomed ominous indeed. However, the continuation of the cold war together with the pent-up demand for consumer goods took up the expected slack and economic activity continued relatively undiminished.

In peacetime the imperative of creating jobs is advanced to justify almost any scheme. It was certainly pushed to support the staging of the Olympic Games in Montreal in 1977. One study that preceded the games came to the natural yet absurd conclusion that the economic impact of the Olympic Games in Montreal depended on how much was spent on them. The more spent, the greater the number of jobs created. A reductio ad absurdum is tempting here. If the Olympic Games create jobs in the construction industry, why not have them every year? Better still, why not have the Olympic Stadium self-destruct every September to be rebuilt every April?

The squander society advocates not only unproductive endeavors but also productive endeavors done in the most inefficient possible way. It is reputed that Sukarno, the Indonesian leader of the sixties, was impressed with a view of economic development as being equivalent to moving masses of dirt around. When confronted with the choice of technologies to build a road he opted for men-with-shovels instead of tractors in order to "create jobs." Upon which, a sharp journalist inquired pointedly why he had not chosen men-with-teaspoons to build the road, since the employment opportunities would be even greater. In general, the more inefficient the technology, the more people a particular job will require.

The employment ethic is a legacy of Keynes. Yet, to use a well-worn cliché, if he were alive today he probably would not be a Keynesian—at least as far as "creating jobs" is concerned. Writing in the 1930s at the depth of the Great Depression, he argued for government spending to stimulate a lazy but wealthy economy to start moving. "Spend more, consume more, save less," was the motto. Keynes praised the ancient practice of pyramid building and the medieval practice of cathedral building as being antidotes to depression. To put purchasing power in the hands of consumers, Keynes even argued for the ultimate unproductive employment of them all: paying people to dig holes and fill them in again.

Today's conditions are not those of the 1930s. The overproduction of the earlier period has been replaced by the increasing resource scarcities of today. The lazy economy of the Depression, peopled by nonspenders and manic savers, is now superseded by a society of spendthrifts, credit buyers, and resulting inflation. A conserver society would have been anathema in the halls of Cambridge when Keynes discussed his theories in 1935. In 1977, Keynes, being the genius that he was, would quite possibly have formulated the problem in different terms. But the contemporary Keynesians, which we all are in some sense or other, have not. We continue to assume that the main problem is that people are not spending enough.

Let us rephrase the employment question. The confusion arises in not distinguishing between the separate aspects of the labor contract. There are four potential transactions involved when a worker is paid by an employer: (1) the worker provides a *productive service* to the employer; (2) the worker may also derive *enjoyment* from the work; (3) the employer pays a *wage* to the worker; (4) the employer may also derive *enjoyment* from employing the worker.

1, 2, and 3 should be separated but most often are not. If the idea is to provide a wage to a worker, independently of whether or not he or she performs a productive service, then this is a transfer payment, which might as well take the form of negative income tax, a gift, or a guaranteed minimum income. The masquerade of pretending to provide a useful job is unnecessary.

If the idea is to provide *job satisfaction* to the worker, then this has nothing to do with 1, unless by sheer coincidence. The rules of efficiency are not necessarily identical with those of pleasure. A factory with specialized repetitive jobs, à la Charlie Chaplin in *Modern Times,* may be efficient without being pleasurable. A nonfactory system with very small production units and with a schedule frequently interrupted by coffee breaks, fooling around, and idle chatter may be more pleasurable and less efficient. If the work is a labor of love then it should be treated as a consumption good, not as an input.

The fact of life in modern times is the progressive substitution of machines for people, because machines can do many tasks better and less expensively. This trend has been observable since the early Industrial Revolution in Britain in the nineteenth century. The Luddites of old tried to resist this trend and destroy machines. The Neo-Luddites attack automation in favor of labor-intensive occupations, which, in most fields, are no longer efficient. When the labor union of the Paris subway kept up pressure for long years to prevent the introduction of automatic ticket punchers, they were clearly opting against efficiency. To condemn an able-bodied person in his or her prime to spending an active life punching holes in tickets is hardly making full use of the human resource. Similarly, when postal workers oppose zip codes and prefer sorting mail by hand, they are resisting efficiency and opting for waste.

It is in the blurring of the distinctions among security as an objective, a wage as an objective, job satisfaction as an objective, and productivity to society as an objective that the present confusion arises which makes the squander society surprisingly appealing to those who have not sorted out these issues.

When a recent OECD study suggests that the labor force in

industrial countries is becoming increasingly redundant (an important fraction of white-collar and even blue-collar workers could stay home all week without significant decrease in total output), then we can choose one of two ways out: either reemploy these redundant workers elsewhere or, if they are unneeded, give them a free income anyway without requiring them to pretend to work, or creating fictitious jobs for them and in the process squandering the human resource.

The squander society advocates creating jobs even if these jobs are fictitious, whereas the conserver society advocates getting the job done with as much economy as possible, including economy of human resource. Given these characterizations, is there in fact a case for the squander society? In our minds there could be one only in either of two circumstances that certainly do not correspond to present reality. If in fact waste were the cost of freedom (which it is not), then the lesser of two evils might be to accept such waste. Second, if an overabundance of materials and a sluggish economy were to combine to create a stagnation as in the 1930s, "spend more, save less" might again be the lesser of two evils. In the absence of either of these circumstances and in the presence of imminent scarcities and environmental deterioration, the squander society is totally unjustifiable and detrimental to social and environmental peace. The squander society would merely be a bad joke if it were not one of the directions toward which our present system is leading. Far from a bad joke it might become a sad reality.

VI
ASSESSING THE
OPTIONS

Well-Chosen Words from Solo Seleccione

"Last letter," Sammy thought. This one should be interesting. Solo Seleccione had come from Italy at an early age and gone to the same high school as Sammy. Solo, however, had gone on to an English literature degree followed by a Master of Fine Arts at Yale, instead of following the MBA route like Sammy. A strange, intense man, Solo would probably have been a better match for Rita Righteous than Angus McThrift. Solo was an artist, an anti-establishment playwright and theater director. He lived frugally and was a faithful conserver of resources—except in the artistic and emotional fields. In fact, he had first met the group when he joined Rita's old cooperative housing project, but had been too uncooperative and temperamental to live with others and had been asked to leave. In matters which inspired his artistic passion his extravagance had no bounds. At present directing a play in the Oregon Shakespeare Festival, Solo had this to say:

> A Forest in Oregon
> July 1st

My Dear Sammy,

I take time off from this beautiful forest where "we feel not the penalty of Adam," the winter's cold, the summer's heat to send you this epistle. I was melancholy the first few days of my stay here. I find actors less and less interesting as a group. Too full of themselves, too artificial. They try to substitute technique (one hand up says "I'm angry," two say "I don't care") for gut feeling. I am living in a tiny cabin on the small salary I am getting from the festival. I'm not yet a member of Actors' Equity and am doing this thing for almost nothing, $45 a week. I eat spaghetti on Monday, Wednesday, Friday and chili the other days. My two pairs of jeans are wearing out—but so what? I never needed more. Thinking of giving one away to the Salvation Army. I have very little use for these material things anyway. I cannot understand what anyone with a $20,000 income can possibly find to do with it.

After being melancholy for a few days, I am extremely excited at the prospect of our next production, which we open in one week. We are doing Henry V outdoors and using terrific sets. I am reminded of the opening prologue "O, for a Muse of Fire, a Kingdom for a stage, princes to act And monarchs to behold the swelling scene." Well, our Muse of Fire is here and our stage is this magnificent forest. We are doing a full happening-audience-involvement production. We asked and obtained a budget of $100,000. I am importing the best sound equipment and lighting from Germany. Christian Fiore will do the costumes and we will have special armor for the battle scenes. Doing the battle of Agincourt in the gorgeous landscape of the state of Oregon will be doing justice to the Bard. I will spare no expense to recreate the total experience of sound, sense, and meaning combined together. It will make Peter Brooks's production of Marat/Sade sound like a play reading.

After the summer I am scheduled to do a TV commercial in L.A. For a week's work I'll get $2,000–to be put toward a rare first edition of Titus Andronicus. After that, I'll come back to the city, find a cheap room, and maybe live on welfare for a while. I'll see you then, old friend.

                                                            Solo

21. THE BIG ROCK CANDY MOUNTAIN VS. THE CONSERVER SOCIETY

This book has outlined five growth scenarios for the future:

The *status quo* or CS_0 was characterized as "doing more with more," implying indiscriminate, haphazard, and undirected growth.

CS_1, or the growth-with-conservation scenario, has "doing more with less" as its guideline. A certain number of strategies were outlined to show how in fact this more efficient, more "economical" growth (in the etymological sense of economical) could be achieved with minimum value change.

CS_2, or the affluent stable state, is an option in which industrial growth is arrested at a certain high level, and once that level is reached energies are diverted to other endeavors. Guided by the motto "do the same with less" it invites us first to define the "same" and then to show how that same can be achieved with less. CS_2 includes all those policies of CS_1 that are compatible with it, in addition to its own. At the practical level CS_2 features many Z's: zero artificial needs growth (ZANG), zero industrial growth (ZIG), zero urban growth (ZUG), etc. The most important Z is the freeze in the growth of artificial needs beyond what we have now.

CS_3 or the connoisseur's conserver society, is the most advanced of the three. It features all the compatible policies of the other two plus a few N's of its own: NANG (negative artificial needs growth), NIG (negative industrial growth), etc. It is a direct challenge to the Big Rock Candy Mountain paradigm of CS_0 and enjoins us to give up Big Rock Candy Mountain climbing in favor of other pursuits. "Do less with less" is its motto and "do something else." It is the genuine postindustrial model rejecting high throughput altogether.

CS_{-1} is the Big Rock Candy Mountain maximized. It is a splurge, a wild party that would make Omar Khayyam's *Rubaiyat* look ascetic by comparison. A short distance from our present society, it leads to "doing less with more" in a vortex of perpetual agitation, which is confused with productive activities.

Before assessing the options we should bear in mind two things:

First, let us reiterate that these are *options*. Although we are obviously in favor of one or another of the conserver options, the Conserver Society Movement must not be seen as a new religion with its missionaries, its creed, and its high priests. Rather, it may be visualized as a potential better idea.

Second, it must be realized that the five options are not so clear-cut as presented in the text. Every intellectual exercise is, of necessity, a simplification of complex reality in order to bring out a fruitful polarization of ideas. Thus, although we continue to contend that the status quo in the United States and Canada is dominated by a philosophy of undirected haphazard growth of the "do more with more" variety, there are nevertheless pockets of each of the other four scenarios coexisting with one another. There are sectors of incredible waste found side by side with models of efficiency à la CS_1, de facto stable states à la CS_2, and low-throughput mini-societies à la CS_3. The relevance of the growth options as presented here relates to what is to be the *dominant* paradigm. Will we have *primarily* a CS_1 society or a CS_2 or a CS_3, etc.? The change will probably not take the form of a solemn announcement by the President of the United States

BIG ROCK CANDY MOUNTAIN VS. CONSERVER SOCIETY 255

and/or the Prime Minister of Canada that they are now decreeing that such and such a society will prevail. Rather the change, if it takes place at all, will be incremental and gradual, interspersed with structural alterations. It will not happen overnight. Futurizing the present is a process not a terminal result because, after all, the future never really comes. We are always in the present.

The five options for the future could be submitted to an assessment involving one, fifty, or a thousand criteria. The choice of assessment criteria and the relative weights we would assign to each involve considerable value judgments. For purposes of simplicity we limit our criteria to ten and invite you to give your own weights and your own judgment on whether a particular criterion is in fact good or bad. The table on pages 256 and 257 summarizes our own assessment of the options.

Criterion 1: Growth Philosophy. CS_0 and CS_{-1} advocate indiscriminate growth, CS_1 efficient growth, and CS_2 and CS_3 high-level and low-level stable states respectively, as far as the throughput dimension is concerned. None of the models advocates an end to personal growth, or to the quest for self-fulfillment or anything of the sort. What changes from one option to the other is the form and nature of growth, but all accept the basic (probably irrefutable) premise that life is motion, and motion implies growth in some things and atrophy in others. The distinction between the models is in their attitudes toward throughput.

Criterion 2: Expected Conservationist Effect. The least conservationist option is, of course, the squander society, whereas the most conservationist is CS_3 with the other three progressively moving toward CS_3. An attempt to translate the conservationist effect in terms of total energy-consumption figures could yield the following data (advanced for illustrative purposes and not as proofs of holy writ).

CS_{-1} would lead to a 5–10 percent growth rate per year in total energy consumption

Assessing the Options

Assessment Criteria	The Five Options				
	CS_0	CS_1	CS_2	CS_3	CS_{-1}
1. *Growth philosophy*	Haphazard	Growth with conservation	Affluent stable state	Ascetic	Very haphazard growth
2. *Expected conservationist effect*	Low	High	Higher	Highest	Lowest
3. *Distance from present value system*	Zero	Very small	Further	Furthest	Very close
4. *Impact on inflation*	High	Short term high, long term low	Low inflation	Very low inflation	Very high
5. *Impact on unemployment*	High *unemployment*	More efficient use of labor	Same as CS_1	High *employment*	Very high *unemployment*

6. *International implications*	Very competitive international relations	Depends on what other countries do	More self-sufficiency	Highest self-sufficiency	Destabilizes international relations
7. *Is the option egalitarian?*	No	Yes, because of rental schemes	Yes	Yes, very egalitarian	Not at all
8. *Does it lead to more statism?*	No (by definition)	More efficient mixed economy	Yes	No, less statism than CS_0	Less Statism
9. *Is the option attractive to the Third World?*	Unfortunately yes	Could be attractive if properly understood	Unfortunately no	Unfortunately no	Unfortunately yes
10. *Expected feasibility in time*	Does not apply	Within the next generation	Within a generation	Within a generation	May become fact right now

CS_0 would lead to a yearly growth rate around the 5 percent mark.

CS_1 would reduce the rate to 2–4 percent, depending on the degree of implementation of its policies. With generalized renting, which could lead to substantial drops in production (without, it is reminded, lowering the standard of living), energy growth could possibly be kept even under 2 percent.

CS_2 would match the energy growth to the rate of the population growth, maintaining in effect zero *per-capita* energy growth. If population does not grow, that total energy consumption will be constant.

CS_3 would allow for actual reductions in total energy used as artificial needs would be phased out and replaced by conserver needs.

The expected conservationist effect is of course relevant insofar as there is a need to conserve. Given today's levels of consumption, it would appear to us that the CS_1 model would probably be sufficient. However, if disasters strike (shutting off of Middle East oil, climate change) then a different assessment would have to be made.

Criterion 3: Distance from present value system. We assume here that the greater the difference between the proposed option's value system and the present value system, the less feasible the option (although some because they disagree with our current life-style would give high marks for great distance from present values, irrespective of feasibility). Under this criterion CS_1 is close to the present CS_0, but then so is CS_{-1}. CS_1 does things efficiently and CS_{-1} does things inefficiently within the same value system.

CS_2 implies some value changes that will turn off some people and turn on others. The turn-offs will generally be connected with a Pavlovian reaction that some people have toward "zero growth." This rejection is absurd unless qualified; it is

nevertheless real. Moreover, some people still will react negatively to any attempt to make a distinction between real and
artificial needs and will come up with the tired cliché "I don't
want anyone telling me some of my needs are more natural than
others." The chapter on ZANG should convince the reader that:
(a) a distinction is possible and (b) it need not have any pejora-
tive or hierarchical intent, since some artificial wants are more
intense than natural ones and therefore demand prior satisfaction. Nevertheless, the Pavlovian reaction based on connota-
tions rather than denotations will still be there.

It is worthwhile noting in passing that strong currents seem to
exist today in favor of the most extreme conserver society's
(CS_3) value system. The success of Schumacher's *Small Is
Beautiful*, which bears a family resemblance to CS_3, and the
proliferation of communities that seek peace with nature presage
well for CS_3. CS_3 is true postindustrialism and it is reasonable to
assume that most advanced regions of the world will adopt that
life-style before the others. In a fit of visionary semi-
facetiousness a noted sociologist suggested that in the year 2000
the world center for Buddhist studies would probably be the
University of California, whereas the center of industrial studies
would be the University of Calcutta. Today, CS_3 is not in the
mainstream, but, given rapid value change and the emergence
of new generations that have not shared the anxieties of the
present ones, what is marginal today might well be mainstream
tomorrow.

Conversely, very strong currents exist in favor of CS_{-1}, especially in nouveaux-riches countries and regions. The Arab emi-
rates, Iran, Saudi Arabia, and all suddenly wealthy nations seem
to compete in extravagance and squandering. Nouveau-riche
Alberta is viewed by some as following the same pattern. According to the Canadian magazine *Maclean's* (April 13, 1977),
Alberta views itself as Camelot West, the "new kid on the
block." It seems to have the Big Rock Candy Mountain philos-

ophy as official dogma. To the question "What does Alberta want?," the answer apparently is "More, more of everything." Retired oil-company presidents drive steer-horned Cadillacs emblazoned with 700 silver dollars. Gold faucets are de rigueur in establishment bathrooms together with "his" and "her" tubs, and the number of bathrooms per luxury house—a surprising but durable symbol of North American affluence—is now nine. In some houses there are more bathrooms than bedrooms. One executive defended this by saying: "Every Christmas my grandchildren come and visit for two days, and I can't be bothered to wait to use the facilities."

For another venerable symbol of North American affluence—the size of the steak—Alberta may be number one. In an Edmonton restaurant a giant steak is available for $60 (1977 prices), but if consumed in less than an hour it is free. The size of the steak? Not, 16, not 21, not 30, but a whopping 72 ounces, in other words four and a half pounds of cow flesh to give its consumer a feeling of achievement in life—and possibly a heart attack.

The more generalized picture of North American CS_{-1} seems to center around what one observer has aptly called "R and T." Once the nine bathrooms are in place and the country home, multiple stereos, power boats, and so on have been acquired, how is one to spend one's money? The answer is "R and T"—restaurants and "tripomania." The restaurant absorbs large portions of the luxury budget, complete with excessive waste—such as ordering many courses, picking at each and sending the bulk to the garbage—and is fueled by the perennial expense account. The trip is the other symbol of the squander society, with incredibly high hotel bills and so on. Since both business and pleasure traveling are symbols of success, tripomania is an important symbol of our present society, with its attendant throughput, effluence, and waste.

Criterion 4: Impact on Inflation. The impact on inflation must be qualified by the fact that there is very little agreement as to the causes of inflation in our present society. CS_0 is a high-inflation model, as is CS_{-1}, because squandering overheats the economy and creates a classical demand-pull inflation. CS_1 may, in the short run, be inflationary because of full-cost pricing. If we start paying for the full cost of what we are consuming, the price will go up, since we currently do not account for pollution, environmental deterioration, and the like, in our pricing. In the longer run, however, the CS_1 model should reduce inflation simply by reducing waste, which is certainly one of its causes. Any reduction there will dampen the spiraling increase in prices. CS_2 and CS_3 should reduce inflation for the classical reasons: they reduce demand for high-throughput production and the economy cools off.

This assessment is preliminary and is offered here as only a rough approximation.

Criterion 5: Impact on Unemployment. Insofar as labor is a resource to be economized like petroleum, gold, or capital, the conserver society options will use less labor to achieve the same objective, whereas the wasteful options CS_0 and CS_{-1} will use more. This, however, must be qualified. CS_1 will require less manpower in forward throughput activity (transforming our raw materials into the surrogate world of shoes, ships, and sealing wax) and more in reverse throughput (transforming the ships and sealing wax back into raw materials) and environmental protection. The same is true for CS_2. For CS_3, on the other hand, deindustrialization should mean a return to labor-intensive activity with a corresponding increase in employment.

As was noted earlier, the attitude toward employment will be a key variable. If it is seen as a means to an end and one allocation of time among many, then more leisure stemming from greater efficiency will be welcomed. If employment is viewed as an end in itself, as our ultimate function, then obviously the less there is the worse the assessment. The conclusion is inescapable: to maximize employment we must choose either a very labor-intensive CS_3 or go full out along the squander route.

Criterion 6: International Implications. These are much more difficult to pin down because there are too many variables. Three plausible hypotheses may be considered.

a. If the United States and Canada alone become conserver societies, this may lead to balance-of-payments difficulties because full-cost pricing and environmental-protection regulation may make North American products even less competitive in the world market. On the other hand, the losses due to full-cost pricing may be more than made up by increasing efficiency in overall production.
b. If the rest of the world becomes a conserver society but not the United States and Canada, the opposite is likely: an increased competitive advantage for the "dirty" North American producers but at the same time higher costs from comparative inefficiency.
c. If the entire world becomes a conserver society the international implications may be less dangerous, since everyone, including such low-cost producers as Japan, Taiwan, and Hong Kong, will abide by the same rules of the game.

The question of international implications must be left open for more detailed analysis. Our temporary conclusion is that it would be much better if all countries adopted the conserver ethic since it would not distort the picture of cost competitiveness and would in addition be compatible with the New Economic Order.* Some of the friction between rich and poor countries stems from the enormous waste in the former, which leads them to devour voraciously a disproportionate share of world resources while poorer countries live in misery. The *squander/*

*Expression used by U.N. to denote proposed reforms in the international economic system.

misery contrast is particularly unnerving and would be eliminated by international adoption of the conserver ethic.

Criterion 7: Is the Option Egalitarian? On the issue of egalitarianism we submit that CS_0 and CS_{-1} do not reduce inequalities among people and nations, whereas CS_1 and CS_2 in particular do. CS_1 features the highly egalitarian rental society. CS_2 goes further, striving to bring everyone within an acceptable range of economic inequality by the adoption of the stable-state philosophy, and by providing a possible guaranteed minimum income. CS_3 is less clear because presumably there may be great differences in the degree of affluence of the various communities that would make up its structure. On purely egalitarian grounds CS_2 is the best option.

Criterion 8: Does It Lead to More Statism? CS_0 is the status quo. CS_{-1} leads to less statism at the cost of greater waste. CS_1 need not lead to greater statism; in the optimum-mix economy, state intervention becomes more efficient but not necessarily greater. The public sector regulates its size to maximum efficiency. In some areas this size may be small or even zero; in others it may be large but not necessarily larger than at present.

CS_2 does lead to more statism in order to implement the various "freezes" inherent in the model. Paradoxically, CS_3 is the *least statist* of all options, perhaps even less so than the squander society. When the scale of enterprise is reduced and self-sufficient or almost self-sufficient units are developed, the need for state intervention—which is a direct result of interdependence—is absent. Nor is the government needed to bring about CS_3 in the first place, since our assumption is that if CS_3 is ever to be born it must emerge spontaneously, not be legislated into existence by the state.

Criterion 9: Is the Option Attractive to the Third World? The poor countries should reject the squander society because of the simple fact that they have very little to squander in the first place.

As one Third World leader put it, "My country has a very high multiplier—but unfortunately there is nothing to multiply." Although the poor countries are adopting CS_0 or the more-with-more philosophy of massive industrialization, come what may, we submit they should instead adopt the CS_1 or CS_2 models. Doing more with less is important when you are poor, and the strategies in CS_1 allow the poorer nations to accede to a higher standard of living faster. In particular, the early adoption of time-sharing or rental schemes, an intelligent management of time, and the use of conserver technology must be warmly recommended. The developing countries have a chance to avoid the mistakes of the industrialized by immediately adopting a conservationist ethic.

The CS_2 option is also desirable for developing countries. By limiting the growth of some areas they may allow others to catch up and thus create balanced growth with preset ceilings.

As far as CS_3 is concerned, some developing areas may find it too close to their existing status quo to be attractive. Superficially "postindustrial" looks much like "preindustrial," but in fact there are significant differences. Although both preindustrial and postindustrial societies are marked by absence of high throughput, postindustrial society may be either labor intensive or information intensive or both, whereas preindustrial society is inescapably labor intensive (such as a peasant-, serf-, or plantation-type economy).

For good or for ill, and given the seemingly universal penchant for believing that the grass is greener on the other side, the developing countries are not likely to adopt postindustrialism without going through the industrial trip themselves. For this reason, CS_1 or CS_2 are the most attractive alternatives.

Criterion 10: Expected Feasibility in Time. Although the business of dating future events is hazardous (with the exception of cosmic events like a total eclipse of the sun), we must venture a few guesses as to the time frame of the various options.

The United States and Canada could well reach the limits to throughput within the next decade, as more and more people demand the right to own cars, pollute, and produce ever-increasing amounts of garbage while gobbling up more and more energy. Barring unlikely breakthroughs, some kind of conserver ethic will probably prevail in some sectors. The conserver society is really a process, not a static product, and it is conceivable that a CS_1-type scenario may be achievable within a decade, if the six strategies are implemented. A CS_2 scenario may be feasible within a generation, and that also is probably true for CS_3. As the babies of today become the adults of the year 2000, they may alight into the new millennium with fewer (or different) hangups than the present generations. Today's status symbols may become as irrelevant to prestige seekers as are those of yesterday: wigs, silk handkerchiefs, butlers, and gold snuff boxes.

Sammy Squander Reconsiders the Big Rock Candy Mountain and the Advice of His Friends

Et voilà. Sammy Squander has a picture of the lives of his five friends. (He had not expected to hear from Fiona Fragment. She always took on the personality best suited to her surroundings, and on a Greek island would forget Sammy completely.) All of his friends' lives are imperfect and perhaps even absurd but certainly more fulfilling than the "prestige" apartment, the phony food, and the junky car which had all come to be symbolized in Sammy's mind by the Big Rock Candy Mountain. In a fit of reforming zeal (Sammy periodically made resolutions which he did not keep) our hero runs to his closet and proceeds to throw out the ridiculous electric back scratcher he bought. It consumed energy and materials and was utterly stupid. Sammy thinks he will change his life-style. Tomorrow he will cancel his appointment with Dr. Fraudoong, who keeps counseling him to adjust to the world he lives in. He wants a new life and, as a first step, he is going to seek advice from Madame Sosostris, the famous soothsayer and crystal-ball gazer, who has just arrived from Europe.

Sammy needs a new future, a complete break with the present, and wonders if his friends' solutions can help him. Angus McThrift certainly does things well but spends a lot of time calculating how, and Lenny Lease's life has no stability. Mr. Middleton seems to carry moderation to extremes, and Rita Righteous's trip might be hard to take, even for her. Solo Seleccione is probably the exceptional "man with a passion." But they all seem to be doing well in their very different ways; maybe to blend their ideas is Sammy's answer and maybe Madame Sosostris can tell him how.

22. A FINAL COMMENT ON MOUNTAIN CLIMBING

What then is the bottom line (or at least our perception of it)? Should we continue the climb toward the elusive bliss point of the Big Rock Candy Mountain like Camus's Sisyphus who, although recognizing the futility of the climb, is fulfilled by it? Should we take short cuts to the top, should we go the slow way, or should we give up mountain climbing altogether?

To encapsulate, we have the following choices. The status quo is the "American way," it is *Homo faber*, Man the Maker, the Transformer, the Throughput Maximizer coercing and disciplining nature to satisfy his needs. The cornucopian extreme of the squander society is like a Roman orgy, laying waste the countryside at an increasing tempo, creating agitation and ending with the wasteland. The CS_1 route has been characterized as the "Scotch" way, thrift with thriving, expansion with economy, activity with art. It makes us better Candy Mountain climbers, more perceptive of what is happening, perhaps more distant and sober climbers but climbers nevertheless.

The CS_2 route, the "Greek" option, offers the virtues of moderation, the stress on optima versus maxima and minima, a love of balance, and the courage to decide when enough is enough.

The last option, the "Buddhist" philosophy, calls for renunciation and thereby liberation from the bondage of candy in favor of a different mountain, the mountain of personal instead of material growth. The extrinsic motivation of commodities and need satisfiers is replaced by the intrinsic motivation of self-worth.

American, Roman, Scotch, Greek, Buddhist. Which? To the reader the final choice. If the choice is made on rational grounds, then the existence of the status quo must loom large in the final assessment of the options. The cost of change, what is called variation cost, must be taken into account. This relates to the pain and discomfort involved in the transition from any one value system to another. When this psychological and institutional price is high the expected benefits must also be great. Consequently, like a poker player faced with a difficult hand, we must weigh the consequences of success and failure. The minimax rule in the theory of games leads to its counterpart here: assume that you have made a mistake in choosing either the cornucopian or conserver society. Which mistake is more costly? If you should conserve and later on discover it was unnecessary to do so because the Age of Abundance lay immediately ahead—is that mistake more costly than the opposite one, of choosing cornucopia only to find ecological disaster ahead? The gamble is like Pascal's of old; he had to decide whether he was going to believe in God. He decided that it was safer to believe in God because at worst if God does not exist it would have meant believing in nothing. If God did exist and one refused to believe in him, the consequences were going to be painful.

To translate the problem of the gamble in terms of the question at hand we can borrow from the insurance metaphor. The conserver society can be seen as an insurance policy for the future with three different premiums giving three different coverages. The first plan, CS_1, involves a very low premium and extensive protection. CS_2 gives a higher protection and a correspondingly higher premium. CS_3 gives the highest protection and exacts a premium which not everyone will want to pay.

By choosing CS_1, we probably make the optimum rational

choice. The premium is so low as to be negligible. If the insurance protection turns out to have been unnecessary, so much the better. It is only the sick-minded person who, having paid for expensive fire insurance, is disappointed that the fire has not actually taken place. We pay for protection from catastrophe, not for profit from disaster. Lack of insurance altogether is living dangerously, perhaps with impunity for some time, as long as there is no fire—but once disaster strikes it is too late to cry over spilled milk.

On rational grounds, CS_1 is the obvious answer. On nonrational grounds, CS_1 is the best option if we think in terms of minimizing the distance from the status quo. If we do not like the latter in the first place, then that question does not apply. The emergence of CS_3-type communities in many parts of North America attests to the fact that for some people CS_3 is not just an insurance policy—it is a more fulfilling life-style altogether.

Whatever the option chosen, we must remember that, in spite of its considerable advantages, the conserver society is not a panacea for all ills, will not right all wrongs or make all sick things whole. There still will be sickness and quest for meaning, and the perennial philosophical problems of existence, of teleology, and of essence will remain—probably as long as there are human beings. But the petty unnecessary problems of overpackaging, uncontrolled garbage growth, voracious energy squandering, and the wild goose chase involved in trying to find happiness in a plethora of things will be reduced. Whatever happens, the fault or benefit will not be in our stars but in ourselves.

EPILOGUE

Weekend at Madame Sosostris's

Madame Sosostris, famous clairvoyant, was the wisest woman in Europe but had a bad cold. Intent on bringing her prophetic skills to the New World, she had rented a chalet in the Adirondacks through her business agent, Moriarty Eugenides, a multinational sort of chap specializing in the import and export of ideas. For this winter weekend, Mr. Eugenides had organized a consultation for Sammy and his friends. They were to spend the weekend meditating, doing sensitivity training and relaxation exercises, and pondering their life-styles. The crowning moment would then be Madame Sosostris's pronouncement of the things to come for Sammy and his friends. Mr. Eugenides had counseled great prudence in the actual clairvoyance. Rapid change had the nasty habit of making predictions so much more difficult now. But his advice was not really needed. The clairvoyant was versed in the arts of ambiguity and ambivalence in the best Delphic tradition. She also had the intuitive sense of the hypothetical syllogism "If you do this, that will result, but if you do not, then that will not result, etc."

Sammy Squander arrived in his Stratobird 2 + 2 with his longtime girl friend, Fiona Fragment, and with Angus McThrift. Fiona had been reluctant to make the trip because of the difficulty of packing, but Sammy, who was anxious for her to meet Madame Sosostris, had offered to help. The problem was that Fiona was afraid to leave home without the full range of her beauty and health aids and a trunkful of clothes. For her head and face alone, she required two wigs, a hair dryer, two shampoos (one for dry hair, the other for oily, to adapt to changing conditions), a facial sauna, an electric toothbrush, dental floss, besides face creams to moisturize, rejuvenate, smooth, nourish, emulsify, and enrich. These were the basic necessities but neither could she imagine wearing any of the ten outfits in her wardrobe suitcase without the appropriate makeup. There really seemed no point in spending three hours and $300 in consultation with the aestheticien Visagiste if she could not always have the Face of the Moment. Fiona was equally careful about the rest of her body—right down to the special astringent-deodorant for her feet. These accouterments had become essential to Fiona's physical and mental wellbeing, indeed, parts of her very self. Yet she was still uneasy, constantly anxious in case the market should produce some new improved thing without her knowledge.

Sammy Squander came similarly equipped and had it not been for the rather large trunk of the Stratobird and Angus McThrift's small efficient overnight bag they would have been obliged to come in a station wagon.

Lenny Lease arrived at Madame Sosostris's straight from the airport where he had flown in from California. He had rented a $39 "weekend-special" car with no mileage charge. Just in case, he had also rented a pair of cross-country skis for $5.25.

Rita Righteous and Solo Seleccione arrived together, wearing army surplus boots, once-blue jeans, and heavy duffel coats from the Salvation Army. They had hitchhiked in from the city. Solo's small used car unfortunately would not start . . . something to do with a frozen fuel pump. Mr. Middleton had been unable to attend but had asked Sammy for a tape recording of the counseling session.

The weekend progressed smoothly. On Friday the reunited friends laughed and joked about old times. Although they had moved along different paths they were still close to one another. On Saturday, they exchanged tokens of affection. Sammy and Fiona gave Angus an expensive suède jacket, Solo and Rita a sterling silver liqueur set from Stephany in New York (Solo wondered when on earth they would use it), and Lenny a beautiful leather wallet (to hold all his rental cards). Angus, a part-time painter, gave one of his recent creations to each of his friends. Rita's gifts came from the store of her own unused possessions: a necklace for trinket-happy Fiona, a soapstone carving for Sammy, a handwoven unisex sweater for Lenny. Lenny gave nothing. Instead, he lent various possessions to each of his friends, enjoining them to return all the items after use.

At dawn on Sunday, the beginning of the day, the beginning of the week, and perhaps the beginning of a new era, Madame Sosostris appeared in all her fullness to the young people. And her fullness was considerable. She had plenished and replenished her body with assorted delicacies from all over the world. Not unlike a statue of the Buddha, she had an imposing presence that lent weight to her wisdom.

In the solemn atmosphere of the inner chamber, Madame Sosostris sat crosslegged on a large cushion, her eyes gazing through the window into emptiness beyond. Cautiously and with deference, Mr. Eugenides approached her with the first question, from Angus McThrift. It read, "My philosophy is efficiency and my passion is to do more with less. Can there be anything wrong with this?"

Madame Sosostris concentrated her immense energy on the fixed point in the very center of Angus's forehead. Her personality seemed to falter, fade, and then gradually be replaced by a new power; the calm, contained, masculine tones of a seventeenth-century Irish churchman, speaking swiftly, came forth:

> "I have a modest proposal for efficiently using this 'prodigious number of children' which in the present deplorable state of the kingdom is a very great additional grievance. . . . I have already computed the charge of nursing a beggar's child . . . to be about two shillings per annum, rags included; and I believe no gentleman would repine to give ten shillings for the carcass of a good fat child, which as I have said will make four dishes of excellent nutritive meat . . . the mother will have eight shillings neat profit. . . . Those who are more thrifty (as I must confess the times require) may flay the carcass: the skin of which, artificially dressed, will make admirable gloves."

It was now Lenny Lease's turn. Approaching with some diffidence he transmitted his question through Eugenides: "My philosophy is to use without abusing. I share by renting. I am a steward not an owner. Can there be anything wrong with that?"

> "The quality of sharing is twice blest. It blesseth him who gives and him who takes. It is the Act of Community that enno-

bles us all. Beware of it nevertheless, for it too can trap you. A man with no Property is a man with no Properties. Own your body and let no one take it. By these few precepts keep thy character: In matters of the flesh neither a lender nor a borrower be, for lending dulls the edge of husbandry. . . ."

Rita Righteous now made her declaration: "Madame Sosostris, I live by the Four Noble Truths. Can there be anything wrong with that?"

"The Four Noble Truths embrace all truth except that which is contained in the Fifth Noble Truth, which is called the 'Ultimate Revelation.' For each of the Four Noble Truths there is a type of satisfaction for our desires: the earth, primary; the fruits of the earth, secondary; the things seen, the tertiary; and the things unseen, the quaternary. These may be ordered into three illustrious scenarios if, first, the duality of truth in our needs and falsity in our yearnings can be balanced by the unique Principle of Selection from the Four Noble Truths" (here the lady's voice rose to a sprightly singing key), "three French hens, two turtle doves, and a partridge in a pear tree."

Returning to her former sober tones she declared:

"The Fifth and Noblest Truth, which is the Final Wisdom of this Sixth Period of the Seventh Millenium, is that statistics prove 'everything' and 'nothing' which, in the eighth age, we will understand to be one and the same."

The last question was from Sammy and Fiona. For the climax climactorum, a clear unambiguous simple query: "Why do we feel so guilty and afraid when we are just trying to be happy?"

Madame Sosostris turned away and once more gazed quizzically beyond. Then with oracular finality she uttered her epilogue:

"Humanity is not failing but, in this land, humans are."

She nodded thrice to emphasize:

"Yes, Humans are failing because ten commandments are too numerous to remember yet they are not numerous enough. They do not command you to look over your shoulder. The mountain which you, with great industry, have been constantly building, destroying, and rebuilding has raised behind you another great mountain—the Mountain of Miseries. The one you see clearly is the Big Rock Candy Mountain, also known as the Gross National Waste. On it you can discern the visible and the invisible—the discarded cans, bottles, paper, the metals and the oxides, monoxides, dioxides, the dying birds, the dead fish, the silent spring, the oiled water and the dyed sand. There, also, are the lost friendships, the abandoned kinships, and the wasted wisdom of venerable age. The waste is mountainous and the list endless.

"I am therefore instructed to inform you that since ten commandments are too many and complicated for you, there are now only three. Mark them well, for they are the minimum, if not sufficient, conditions for a good life:

"Thou shalt not waste heedlessly anything, visible or invisible.

"Thou shalt not destroy the fruits of nature, your own endowment or the other creations of the Almighty.

"And this above all:

"Thou shalt look to the consequences of all your acts as far in the future as you can see."

As Sammy and his friends retreated in awe of the Great Lady, her features broke their solemnity and the faintest trace of a contented smile dawned upon the countenance. Much wisdom had been vouchsafed them in so little time.

THE GAMMA
RESEARCH TEAM

Project Director: Kimon Valaskakis
(Professor of Economics, Université de
Montréal, and Director, GAMMA)

Assistant Directors: Peter S. Sindell
(Anthropologist, formerly McGill University, and
Senior Research Associate, GAMMA)
J. Graham Smith
(Professor of Management, McGill University, and
Associate Director, GAMMA)

Research Associates: Laurent Amyot
(Professor, Institut de Génie Nucléaire, Ecole
Polytechnique de Montréal)
Paris Arnopoulos
(Associate Professor of Political Science, Concordia University)
Symon Chodak
(Professor of Sociology, Concordia University)
Max Dunbar
(Professor, Marine Sciences Centre, McGill University)
W. Lambert Gardiner
(Psychologist, formerly Concordia University)
Jacques Henripin
(Professor of Demography, Université de Montréal)
Roland Jouandet-Bernadat
(Associate Professor, Ecole des Hautes Etudes Commerciales)
Iris Fitzpatrick-Martin
(Research Associate, GAMMA)
Franz Oppacher
(Assistant Professor of Philosophy, Concordia University)
Eugen Scanteie
(Economist, Université de Montréal)

Stanley Shapiro
   (Professor of Marketing, McGill University)
Irene Spry
   (Emeritus Professor of Economics, University of Ottawa)
Benno Warkentin
   (Professor of Soil Science, Macdonald College, McGill
   University)

INDEX

THE
CONSERVER
SOCIETY

VALASKAKIS • SINDELL
SMITH • FITZPATRICK-MARTIN

Do we have a future? If the answer is yes—and the authors think it is—then what kind of future do we really want? What kind of future will we actually provide for ourselves?

This study, sponsored by two universities and fourteen government agencies, imagines five possible scenarios—doing more with less, doing more with more, doing the same with less, doing less with less, and doing less with more—and concentrates on the most likely one. We all want to eliminate waste, develop and expand the economy in harmony with rather than in opposition to the ecological system, and conserve physical and human resources.